Fire
Investigation

Forensic Science Series

Edited by James Robertson
Forensic Sciences Division, Australian Federal Police

Firearms, the Law and Forensic Ballistics
T A Warlow
ISBN 0 7484 0432 5
1996

Scientific Examination of Documents: methods and techniques, 2nd edition
D Ellen
ISBN 0 7484 0580 1
1997

Forensic Investigation of Explosions
A Beveridge
ISBN 0 7484 0565 8
1998

Forensic Examination of Human Hair
J Robertson
ISBN 0 7484 0567 4
1999

Forensic Examination of Fibres, 2nd edition
J Robertson and M Grieve
ISBN 0 7484 0816 9
1999

Forensic Examination of Glass and Paint: analysis and interpretation
B Caddy
ISBN 0 7484 0579 9
2001

Forensic Speaker Identification
P Rose
ISBN 0 415 27182 7
2002

The Practice of Crime Scene Investigation
J Horswell
ISBN 0 7484 0609 3
2003

Fire Investigation
N Nic Daéid
ISBN 0 415 24891 4
2004

Fire
Investigation

EDITED BY
NIAMH NIC DAÉID

CRC PRESS

Boca Raton   London   New York   Washington, D.C.

Library of Congress Cataloging-in-Publication Data

Fire investigation / edited by Niamh Nic Daéid.
      p. cm.
    Includes bibliographical references and index.
    ISBN 0-415-24891-4 (pbk.)
    1. Fire investigation. I. Daéid, Niamh Nic, 1967-

TH9180.F477 2004
363.37′65—dc21   2003056517

Visit the CRC Press Web site at www.crcpress.com

Contents

Contributors

Dr Niamh Nic Daéid
Forensic Science Unit
University of Strathclyde
Royal College
Glasgow, Scotland

Dr Caroline Maguire
Mac Daéid and Associates
Consulting Scientists
Co. Meath, Ireland

Reta Newman
Pinellas County Forensic Laboratory
Largo, FL, USA

Martin Shipp
FRS Building Research Establishment
Watford, UK

Eric Stauffer
MME Laboratories
Forensic Service Unit
Suwanee, GA, USA

Dr John D. Twibell
Department of Forensic Science and
   Chemistry
Anglia Polytechnic University
Cambridge, UK

Preface

As with many compilations of work such as this, the time required to pull everything together is always underestimated and this volume has taken somewhat longer than expected to finally appear. All of the authors are extremely busy individuals and it is very much to their credit that they have produced, sometimes at relatively short notice, the very high quality of material that appears within this text.

I hope that this volume will act as a valued document for both practising fire scene investigators and laboratory scientists involved in the analysis of debris recovered from the fire scene, as well as being a general reference text which is informative to those wishing to study the field. The text provides a very basic initial introduction to some of the concepts involved in the phenomena of fire followed by more detailed aspects of fire investigation, specifically those involving non flammable liquid initiated fires. Scene reconstruction and computer modelling is also comprehensively dealt with. The final chapters involve an examination of laboratory based analysis of debris recovered from scenes and their interpretation in light of matrix products which may also be present in a sample.

I would like to acknowledge all the hard work and patience of the contributing authors and ask them for their forgiveness for the necessary nagging and cajoling which the production of this work inevitably involved. I hope they agree that the end result has been worthy of their efforts.

Finally, I would like to dedicate this work to the memory of Diarmuid Mac Daéid who sadly passed away before it was completed.

Niamh Nic Daéid
Glasgow

An introduction to fires and fire investigation

Niamh Nic Daéid

Introduction

Fire investigation can occur in two different stages. The first involves examination of the fire scene to determine the cause, origin and development/spread of fire. The second involves laboratory analysis of samples recovered from a fire scene normally when arson is suspected. While both of these may be linked together, they may be activities carried out by different personnel with different backgrounds and experience.

Scene investigation

In order to successfully carry out fire scene investigations, the investigator should have an understanding of a variety of concepts. These include:

- the fundamental practices and methodology involved in fire scene and crime scene investigation;
- the necessary conditions for a fire to be initiated and maintained;
- knowledge of the dynamics of a fire and factors influencing fire development and spread;
- knowledge of different types of fuel packages, their auto-ignition temperatures, behaviour in fires and the level of heat release which they may produce;
- different types of burn and smoke patterns and their interpretation;
- sampling protocols, packaging, etc.

Only with a sound knowledge of these and other factors can an investigator carry out his or her scene investigation efficiently and correctly.

Laboratory analysis

Laboratory based analysis requires the appropriate skill and knowledge of relevant scientific instrumentation, proper laboratory practice in dealing with crime scene evidence and an understanding of the nature of materials including flammable liquids, their pyrolysis and combustion products as well as an ability to interpret the results of their analysis.

This chapter acts as a summary of the phenomena of combustion, the development of fires in compartments and the spread of fire from one compartment to the next. Reference has been made liberally to texts such as *Kirk's Fire Investigation* [1] and *An Introduction to Fire dynamics* [2] and the reader is referred to these texts for a fuller explanation.

Types of fires

The determination of the type of fire which has occurred often falls to either fire brigade personnel, police officers, scenes of crime officers, forensic scientists or private fire investigators who specialise in fire scene investigation. Deliberate fires or arsons are estimated to account for between 50% and 60% of fires in some parts of the UK [3], with direct financial costs of billions of pounds per annum. Over the last 10 years the number of arson fires in the UK has increased by over 40% and arson fires in vehicles have tripled. The average national detection rate of arson fires remains low at 8% in 2002 (for England and Wales) [4]. Motivations for arson are often varied and complex and can include:

- criminal intent such as covering up other crimes (theft, murder);
- financial gain (insurance claims);
- civil disorder (youth disorder, vandalism);
- malicious intent (grudge/reprisal against a particular race, religion, societal group);
- as part of a series of crimes of a known arsonist;
- acts of terrorism with motivations such as urban unrest, racial or religious hatred or for political reasons.

Not all fires are deliberately set and many arise from various types of accidental cause. This can include spontaneous combustion of materials, careless discarding of smoking materials, careless use of candles, electrical faults and so on. Many of these causes are dealt with in Chapters 2 and 3.

Fire investigators

Because of the differing nature of fire scenes, a number of interested parties can be involved. These may include the fire brigade and brigade fire investigation units, the police and assorted support team (forensic scientists, scenes of crime officers, etc.), insurance agents and loss adjusters, various independent fire investigators, and representatives of relevant local authorities in relation to public health and/or safety. Each team on site may wish to carry out an investigation and prepare a report. In the UK the recent Arson Scoping Study [6] has recommended that agencies such as those mentioned should strive to work co-operatively (where practical) in an effort to attempt to reduce the incidents of fire.

Fire and combustion

Fire can be defined as an exothermic chemical reaction involving the oxidation of some substance (a fuel) resulting in the release of energy in the form of light and heat. The

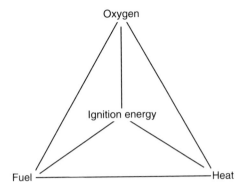

Figure 1.1 The fire quadrangle

conditions necessary for combustion to occur are the presence of a fuel, a source of oxygen (air) and heat. There is also the requirement, in most cases, for an initial ignition source to provide sufficient energy to overcome any energy barriers and ignite the fuel present.

The presence of fuel, oxygen, heat and an appropriate ignition source of sufficient energy comprise the fire diamond or quadrangle (Figure 1.1). Once the fire has been established, fuel, oxygen and heat alone comprise the fire triangle and removal of any of these three conditions results in the suppression of the fire.

Fire is a series of exothermic oxidative reactions involving the fuels present. Most fuels will contain hydrogen and carbon, and water and carbon dioxide are the major products of combustion with other products including carbon monoxide and oxides of sulphur, nitrogen and other compounds which may be present and which contribute to smoke.

Types of Fuel

Most fires will involve combustible solids, though liquid and gaseous fuels are also found. The range of fuels involved is very large and encompasses simple hydrocarbon gases to chemically complex solids, either natural, synthetic or semi-synthetic. These fuels will burn under appropriate conditions, reacting with oxygen and releasing heat and combustion products. Flame itself is a gas phase phenomenon and flaming combustion of solids and liquids require their initial conversion to a gaseous state. For liquids this is relatively straightforward through boiling at the surface. Solids, unless they sublime, must undergo a chemical decomposition process called pyrolysis to produce sufficient low molecular weight volatile components which can enter the gaseous phase. In order to achieve this, a high degree of energy is required and as a consequence the temperature at the surface of burning solids tends to be quite high.

Some relevant properties of materials

Flammability (explosive) limits/range

Mixtures of flammable gases or vapours with air will combust only when they are within particular ranges of gas–air concentration. Outside of these limits the fuel–air mixture is either too lean or too rich to ignite. If the fuel–air mixture is confined in a closed system then the mixture must explode to ignite and the explosive limits and flammability limits are

3

Table 1.1 Flammability limits of common gases and liquids (adapted from reference [2])

Substance	Lower FL% vol in air	Upper FL% vol in air
Hydrogen	4.0	75
Methanol	16.7	36
Acetone	2.6	13
Carbon monoxide	12.5	74.0
Diethyl ether	1.9	36
Paint thinner	~0.8	~6.0
Kerosene	0.7	5.0
Petrol (100 octane)	1.4	7.6

the same. In an open system other factors such as temperature of the surrounding medium will effect the flammability limits. Table 1.1 illustrates some flammability (explosive) limits of common gases and liquids.

Vapour density [2]

This is a property of a vapour that predicts its behaviour when released in air. It is defined as the density of the vapour relative to the density of air and is calculated by dividing the molecular weight of the gas by that of air (approx. 29). A gas with vapour density greater than 1 is heavier than air and will tend to settle through the air into which it is released until it encounters an obstruction, when it will tend to spread out at this level. A similar situation exists with a gas of vapour density less than one which will spread out at ceiling level. Inevitably there will be some mixing of the gas with air due to diffusion, the degree of which will depend on the difference between the vapour density of the gas and that of air (=1). The greater the difference, the less diffusion or mixing of the gases will occur.

Vapour density will cause the formation, over time, of a progressively more or less dense layer of gas in a compartment; moreover, if it has a high vapour density the gas may 'flow' down stairwells or into cellars in a building. It should also be noted that the vapour density and diffusion rate of the lightest (most volatile) component in a mixture is the one that will determine the spread and ignitability of vapours, not the properties of the bulk liquid.

Flash point

The lowest temperature at which liquid fuel produces a flammable vapour. At the flash point the vapour pressure of the fuel is equal to the fuels' lower flammability limit. This flash point is quoted as a closed container measurement (closed cup) or an open container measurement (open cup) (Table 1.2).

Fire point/Flame point

The lowest temperature at which a liquid fuel produces a flammable vapour in sufficient quantity such that if a source of ignition is introduced the vapour will ignite and is usually a few degrees above the flash point.

4

Table 1.2 Flash point and fire point for some common liquids (adapted from references [1] and [2])

Fuel	Open cup (°C)	Closed cup (°C)	Fire point (°C)
Acetone	−20	—	—
n-Decane	46	52	61.5
Methanol	11	11	13.5
p-Xylene	27	31	44
Petroleum ether	−18	—	—
Kerosene	38	—	—
Diesel	52	—	—
Petrol (100 octane)	−36	—	—
Petrol (low octane)	−43	—	—

Table 1.3 AIT values (adapted from references [1] and [2])

Fuel	Temp (°C)
Acetone	465
Ethanol	363
Methanol	385
Petroleum ether	288
Kerosene	210
Petrol (100 octane)	456
Petrol (low octane)	280
Linseed oil (boiled)	206
Polyethylene	488
Polystyrene	573
PVC	507
Polyurethane	456–579
Soft wood	320–350
Hard wood	313–393

Ignition, auto-ignition temperature (AIT)

The temperature at which a fuel will ignite on its own without any additional source of ignition. All fires (with few exceptions) occur because there is a local high temperature in an area in which there is a fuel–air mixture within its flammability range. This area can be very small, but is one in which the ignition temperature of the fuel is reached or exceeded. The attainment of localised high temperatures is not difficult or uncommon over small areas (e.g. striking a match, or creating a spark) and it is the temperature attained that is the important factor. So long as the heat energy can be transferred from the source to the fuel ignition, fire may result (Table 1.3).

Thermal inertia

The ease with which a material can be ignited will depend on how rapidly its surface temperature will rise when exposed to a heat source. Energy is transferred from a heated area

Table 1.4 Thermal inertia values of common materials (adapted from references [1] and [5])

Material	Thermal conductivity (W/m · K)	Density (kg/m^3)	Heat capacity (J/kg · K)	Thermal inertia (WJ/m^{4}K^2)
Copper	387	8940	380	1.31×10^9
Concrete	0.8–1.4	1900–2300	880	1.33×10^6 to 2.02×10^6
Gypsum	0.48	1440	840	5.8×10^5
Oak	0.17	800	2380	3.2×10^5
Pine (yellow)	0.14	640	2850	2.5×10^5
Polyethylene	0.35	940	1900	6.2×10^5
Polystyrene	0.11	1100	1200	1.4×10^5
PVC	0.16	1400	1050	2.3×10^5
Polyurethane	0.034	20	1400	9.5×10^3

of a material to an unheated area of material at a rate dependent on the difference in temperature between the two materials and the physical properties of thermal conductivity (how good the material is at conducting heat energy), density and heat capacity or specific heat (how much heat is necessary to raise the temperature of the material). These three properties give the thermal inertia of the material.

Thermal inertia (WJ/m^{4}K^2) is given approximately by the expression:

Thermal inertia $= k\rho c$

where k is the thermal conductivity (W/m · K); ρ the density (kg/m^3); and c the heat capacity (J/kg · K).

Thermal inertia does not remain static during a fire since once a constant temperature is reached the impact of density and heat capacity become less relative to thermal conductivity. The effect of thermal inertia is most felt in pre-flashover fire situations and has a significant effect on how fast the surface temperature will rise when affected by a heat flux. The lower the thermal inertia, the faster the surface temperature will rise (Table 1.4).

Heat release rate (HRR)

Heat release rate is a measure of the amount of energy a specific type of fuel can contribute to the heat flux in a fire. HHR (expressed normally in kilowatts) is controlled by the chemical and physical properties of the fuel and the surface area of the fuel package. Typical HRR values for different types of furniture are shown in Table 1.5.

Heat transfer [1]

The mechanisms of heat transfer have major importance to fire scene investigation as it is through these mechanisms that a fire can spread from its point of origin to other sources of fuel. There are three processes by which heat may be transferred. In most cases one of the three methods tends to predominate but all contribute.

Table 1.5 Typical HRR values for different types of furnitures (adapted from references [1,2] and [5])

Item	Peak HRR (kW)
Upholstered chair	150–700
Waste paper bin (small 1.5–3 lb)	4–18
Sofa	250–3000
Cotton mattress	40–970
Polyurethane mattress	810–2630
TV set	120–290
Christmas tree	500–650
Latex foam pillow (cotton/polyester)	112
Wardrobe (68 kg, plywood)	3500
Closed curtain (cotton/polyester)	267
Open curtain (cotton/polyester)	303
Wood shelves with video tapes	800
Plastic waste bin (0.63 kg) filled with empty cartons	13
0.61 m^2 pool of petrol	400
1.9 L pool of camping fuel	900–1000
1 bar electric fire	1

Conductive heat transfer occurs by the transmission of molecular vibration. It occurs mainly within a solid item. Heat is transferred by direct contact and the rate of transfer is dependent on factors such as the thermal conductivity of the material and the temperature difference between cooler and warmer areas.

Convective heat transfer involves the transfer of heat by physical movement of materials and occurs only in liquids and gases. Hot gases rise and spread heat to nearby ceilings and walls. This can be one of the primary mechanisms by which heat energy is spread during a fire.

Radiative heat transfer is transfer in the form of electromagnetic energy directly from one object to another (e.g. infra red radiation from the sun). All objects having a temperature above absolute zero will radiate heat. Objects will feel cold if they radiate heat faster then they absorb heat and will feel hot if they radiate heat slower than they absorb heat. In environments where everything is at the same temperature, radiation and absorbance of heat occurs at the same rate. The rate of radiant heat transfer is dependent upon factors such as the temperature of the emitting object/area, the rate of heat release, the temperature of receiving surfaces, the finish and colour of the receiving surface, the angle at which the receiving surface is in respect of the emitting surface. This is one of the main methods of heat transfer before flashover and is a major contributor to fire spread throughout compartments and structures.

Flames are also spread by the direct flame impingement onto nearby surfaces.

Combustion [1,2]

The process of combustion of materials is dealt with in more detail in Chapters 2 and 7. A brief introduction is presented here.

Glowing combustion occurs when solid fuels are not capable of producing sufficient quantities of gas during pyrolysis to sustain a flame. If access to the oxidant (air) is limited, glowing combustion may result. If a smouldering fire is given enough access to the oxidant, the reaction rate may increase, generating more heat possibly giving rise to flaming combustion.

Flaming combustion is the most commonly recognised type of fire encountered and occurs with gaseous fuel sources only. Some of the heat produced by burning is fed back to the fuel, producing more gaseous phase and supporting the flame. The colour of the flame can give some indication of the composition of the fuel. The flame from a pure hydrocarbon gas, mixed with air is blue (e.g. Bunsen burner flame). An orange, yellow, red or white flame is seen when carbon and other solid or liquid by-products of incomplete combustion are present. In solids such as wood, pyrolysis occurs readily under the influence of heat, producing flammable gases which act as a fuel source for combustion. Such fires are often accompanied by the production of charcoal which may undergo glowing combustion. The degree of flame and/or char production depends on the chemical composition of the fuel present, for example, coal contains less volatile components than wood and as a consequence does not produce as much open flame or as much char.

Spontaneous combustion may be described as a process whereby a material self-heats, eventually exceeding its AIT. The process of heat build up may take a considerable amount of time. The driving force behind such combustion is that an exothermic reaction generates heat and if this heat cannot be dispelled it can build up within the fuel mass and raise the temperature. Eventually, the temperature may rise to reach the point of ignition of the entire mass. Some commonly encountered products susceptible to self-heating are drying oils (linseed oil, tung nut oil, fish oils) which harden by oxidation of double bonds in their fatty acid constituents. This reaction generates heat. Auto-oxidation of cotton rags soaked in linseed oil can occur with temperatures of >70°C and within a period as short as 5 hours given the correct conditions [2]. Some enzymatic reactions can be exothermic resulting in self-heating and microorganisms have been known to cause self heating in coal refuse piles and particularly in hay stacks. Spontaneous combustion of different materials is discussed in detail in Chapter 2.

Explosive combustion can occur when vapours, dusts or gases, premixed with the appropriate amount of air, are ignited. Because premixing of the fuel and oxidant occur, the entire combustion process occurs almost instantaneously and are recognised as an explosion. In some cases the first phase of a fire is an explosion, when flammable gases and air have mixed and become ignited.

Development and behaviour of fires in compartments

A fire develops through a number of fairly predictable stages. Initially a source of ignition is required at a site suitable for flaming combustion to occur. The materials begin to burn in a sustained ignition with an open flame which remains once the initial source of ignition is removed. This ignition is localised to the first fuel ignited. The fire plume emits hot gases generating a heat flux. These gases typically containing soot, water vapour, CO_2, SO_2 and other toxic gases. Convection carries these products and heat to the upper parts of the compartment and draws oxygen in at the bottom to sustain combustion. The increasing gas layer at the ceiling radiates heat into the room (Figure 1.2).

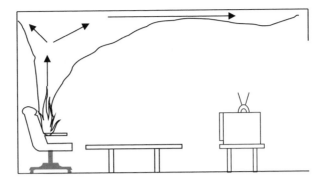

Figure 1.2 Initiation and free burning

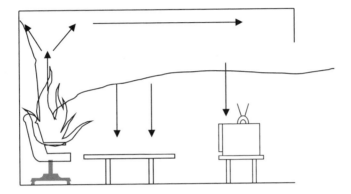

Figure 1.3 Growth period

Growth period

Convection and radiation spread the flames upwards and outwards from the original fuel package until nearby fuels reach their AIT and become involved in the fire. Radiative heat may spread the fire laterally depending on factors such as the proximity of fuel packages to each other. The fire grows by progressively spreading to involve adjacent combustible items. Hot gases composed of toxic gases, partially combusted pyrolysis products, soot and smoke rise to form a fuel rich layer at the ceiling, the temperature of which steadily increases. The lower part of the room will still be rich in oxygen and the rate of burning within the area continues to increase with a consequent increasing release of heat. As the fuel rich gas layer gets lower it may eventually ignite as some of its constituents may reach their AIT or by direct flame contact. This stage is called flameover and involves a rolling flame front within the hot gas layer (Figure 1.3).

Flashover

Even without flameover occurring the hot gas layer is radiating heat into the room. This causes items in the room to progressively heat up and when the layer reaches a temperature

9

of approximetely 600°C it is generating approximately 20 kW/m^2 [2]. In a normally proportioned room this is sufficient to raise the temperature of cellulosic fuels within the room (furniture, carpets, etc.) to their AIT and simultaneously ignite in a process called flashover. Flashover is a transition from a fire involving one fuel package after another to a fire which involves all available fuel in the compartment. At the time of flashover, ventilation in the compartment becomes a restriction on the amount of oxygen available for combustion to occur, and the minimum size fire that can go to flashover in a given room is a function of the ventilation provided through an opening (ventilation factor) (Figure 1.4).

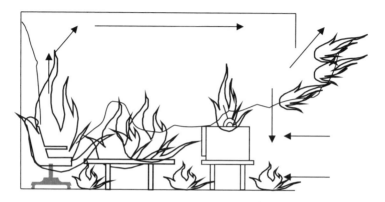

Figure 1.4   Flashover

Post-flashover (steady state burning)

Fire is a balancing act between fuel, heat and air. If the ventilation is limited then the fire will progress at a slower rate involving slower temperature rise and greater production of smoke. Ignition of the smoke layer will take longer or may only occur outside of the compartment if the oxygen supply is limited. If the fuel does not burn fast enough or produce enough heat, flashover may not be reached. Once post-flashover or steady state is reached all involved fuels will continue to burn as long as oxygen is available until the fuel is consumed (Figure 1.5).

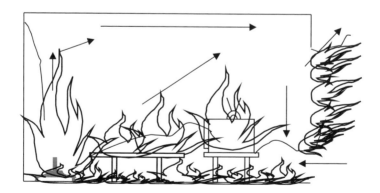

Figure 1.5   Post-flashover

Smouldering

Eventually, as the fuel available becomes exhausted open flaming combustion becomes gradually less and glowing combustion becomes more prevalent. This can also occur if the oxygen levels in a developing fire drops (below *c.*16%). The fuel may still remain in a heated state and the reintroduction of oxygen can cause the fire to re-ignite with explosive speed. Such a scenario is sometimes called backdraft.

Factors affecting fire growth

The rate of growth and development of a fire is very complex and depends on various factors. Obviously, it will depend upon the available area over which combustion can occur. Rate of development also depends upon the nature of the combustible materials, both their construction and geometry, as well as the location of the ignition source. In order to attempt to understand the spread of a fire it is necessary to know something of how specific materials used in modern building construction respond in fire conditions. The structural responses of the building is in general governed by the loading (weighting) which the structure must bear in various places. There are different types of loads which become important under conditions of fire. These include:

- Fire load – the amount of combustible material in the space containing the fire.
- Design load – the amount of material the component is designed to carry.
- Live load – the amount of load the component actually carries.
- Load resistance – the ability of the component to support the load.

As the temperature of a component of a structure increases, its load resistance decreases and collapse can occur once the load resistance is less than or equal to the imposed load. Structural deformity is an example of reduction in load resistance. The speed of fire growth is very dependent not just on the fuel load within a given compartment but also on the amount of heat release (how much and how fast) which occurs from a given fuel package within the compartment and how this heat affects other fuel packages within the room.

Other effects on the time to steady burning include the orientation and direction of the spread of the flame as this can effect the mode of heat transfer from the flame to unaffected fuel. This means that the angle of fuel source relative to the flame will effect how quickly the flame can spread. Vertical upwards propagation can be extremely rapid, raising the temperature of the surrounding area by radiation and convection, very quickly. Examples would be curtains, wall hangings and high stacked storage of combustible items. At the other extreme flame burning vertically downward generally spreads at a slower rate as heat transfer is inefficient.

Generally, the heat transfer required for the ignition of solids is of the order of $10 \, kW/m^2$ which is equivalent to a gas layer temperature of approximately 400°C and the onset of flashover occurs when the hot gas layer imposes radiant energy levels of about $20 \, kW/m^2$ (~600°C). Similar heat transfers are known for the various fire stages of fire growth [2].

Ignition of solids	$10 \, kW/m^2$
Flashover	$20–40 \, kW/m^2$
Developed fires	$50 \, kW/m^2$
Severe fires	$100 \, kW/m^2$

Once flashover has been reached, full room involvement will follow in the majority of cases unless the fire is extinguished. It has been shown that the time to flashover from open flame can be as short as 1–2 minutes in residential fires with contemporary furnishings [7].

The maximum reach of flame is greatly affected by the surrounding space because this in turn affects the amount of available surface of flame into which air can be entrained. Essentially, during steady burning combustion requires a constant amount of entrained air. If the surrounding surfaces impinge on the entrainment of air the flame will grow in height or length to increase air entrainment.

Spread of flame to other compartments

External flaming occurs when burning within a compartment is controlled by the levels of ventilation present. The amount of air is often insufficient to burn all the fuel vapours produced within the confines of the compartment and excess fuel vapours will flow out of the compartment, mix with air and burn. If the fire receives part of its ventilation from an open door within the building, hot combustion products and unburned volatiles will emerge from this opening, heating up fuels in adjoining corridors and compartments. As flames entrain out of the compartment of origin they provide an ignition source for the preheated fuel. In this way fire can spread to the rest of the building through suitable pathways.

Conclusion

The purpose of this chapter was to give a brief and relatively simplistic explanation of how fires happen. It is also intended to equip the reader with some of the physical and chemical values for variables such as vapour density and AIT values which are required in the understanding of fire development and behaviour. Some of the concepts touched upon here will be explored in greater depth elsewhere in this volume.

References

1   J. DeHaan (1997) *Kirk's Fire Investigation*, 4th edn, Prentice Hall, New Jersey.
2   D. Drysdale (1999) *An Introduction to Fire Dynamics*, 2nd edn, John Wiley and Son, New York.
3   Department for Transport (2000) Local Government and the Regions, Fire statistics, United Kingdom, DTLR, London.
4   Arson Control Forum (2002) Leading the Fight Against Arson, the First Annual Report of the Arson Control Forum, London.
5   NFPA 921 (1998) Guide for fire and explosion investigations, National Fire Protection Association, PO box 9101, Quincy, MA 02269-9101, USA.
6   Office of the Deputy Prime Minister (2002) The Burning Issue: Research and Strategies for Reducing Arson, London.
7   N. Nic Daéid (2002) Fire and Explosion Investigation Working Group, Live Burn Project – Cardington, 2001, The Cardington Report – Main Room Burn, Glasgow.

2

Fires from causes other than electrical malfunctions

Theory and case studies

Caroline Maguire

Introduction

Much of the work of the independent forensic investigator deals with incidents that do not necessarily involve the commission of a crime. They do, however, involve injury to persons, property or the environment, which may result in an action before the civil courts or a sworn hearing before an arbitrator. As an independent fire investigator in the Republic of Ireland, I accept instructions both from the insurance industry and from representatives of private clients. My brief is to investigate the causes of fires, and note any circumstances which may have contributed to the severity of injury to persons or damage to property. The insurer needs to know whether a fire was accidental or deliberate; whether insurance policy conditions (warranties) have been fulfilled; whether there has been negligence on the part of their insured or a third party, or contributory negligence on the part of injured parties. Have there been breaches of statutory duty under the civil law, or breaches of planning or bye laws? Has there been compliance with building and fire regulations? Were electrical, gas and oil installations carried out in accordance with recognised standards? Has there been non-disclosure of business carried on or equipment installed in domestic premises? Was all the damage claimed for caused by a particular incident? Was all the property listed actually on the premises? How can such losses be avoided or minimised in the future? The private client looks for evidence to support or defend a claim for damages, or to assist him in a dispute with his insurers.

The investigator must maintain independence and objectivity, while examining, noting, photographing and arranging the preservation of evidence, in the knowledge that every detail may have to be recalled on the witness stand, usually years later. Site and interview notes and photographic negatives must also be preserved; few things lend more weight to testimony than nine-year-old handwritten site notes, complete with original soot and water smudges!

The independent investigator is sometimes at the disadvantage of receiving instructions days, weeks, or sometimes much longer, after an incident. This is particularly likely to be the case where instructions come from representatives of private litigants. There may be little or nothing left of the physical evidence and the scene may have been demolished and even rebuilt. Work has to begin with reports and photographs produced by police, fire brigade and other investigators, and such physical samples or test reports as may be available. Whatever the nature of the incident, the investigation will centre on the scene of the fire, the structures, materials, processes and personnel involved, identification of possible sources of ignition and evidence of eyewitnesses.

This chapter examines, in some detail, various fire causes other than flammable liquid initiated fires and illustrates these with many case examples.

Combustion

Fire is the chemical combination of oxygen with other molecules in an exothermic reaction. A certain threshold energy is required to initiate production of flammable vapours and trigger their combination with oxygen in the process of ignition. This reaction can only occur when the volume of flammable vapours mixed with air lies within a range identified as between the upper and lower limits of flammability for the vapours involved (see Table 2.1). Exothermic energy of combustion is produced as heat and light and the physical products include a range of solid, liquid and gaseous substances. Their composition depends on the chemistry of the original fuel materials, the amount of oxygen available and the temperatures reached during the reaction. The dynamics of combustion of gases, liquids and solids are comprehensively dealt with in publications such as NFPA (1976 and later editions) [1], and Drysdale (1985) [2]. Control of combustion, by the deliberate use of fire, has enabled mankind to develop and enjoy greater use of his environment, whether as hunter/gatherer or as denizen of a modern city.

Combustible materials

Most combustible materials are derived ultimately from vegetable matter, the carbon based, naturally occurring, end-products of photosynthesis. The deliberate burning of combustible materials as fuel for defensive, heating, lighting and cooking purposes is a very ancient part of human history. Following a long dependence on keeping alive the embers of naturally occurring fires, our ancestors developed the skill of igniting suitable materials, by friction and other methods still in use. Dried plant materials, the earliest fuels, are still employed in various guises. Wood, whether as waste or deliberately harvested, is still in worldwide use as a fuel. Fossil fuels have come to prominence since the industrial

Table 2.1 Flammable gases in common use

Gas	Common name/use	Flammable limits (% volume in air)	Ignition temp (°C)	Vapour density
Methane	Natural gas/domestic	5.0–15.0	540	0.6
Butane	Camping gas	1.9–8.5	540	2.0
Propane	LPG/domestic	2.2–9.5	450	1.6
Acetylene	Cutting/welding	2.5–100	305	0.9

revolution and large-scale incineration of waste materials as fuel is becoming acceptable in industrialised societies today. Although combustion and combustible materials have been part of daily life for millennia, accidental fire, brought about by ignorance, carelessness, or negligence, is still an almost daily occurrence in most communities.

Before considering their behaviour in uncontrolled fires, it is useful to examine the chemical and physical structure of carbon-based materials and some of the situations in which they may be encountered.

Cellulose

In photosynthesis, processes in the green plant cell absorb solar energy and combine atmospheric carbon dioxide with water, producing carbohydrate and returning oxygen to the atmosphere. The immediate end-product is D-glucose, a 6-carbon sugar molecule. Some of the D-glucose is quickly oxidised within the plant cell to provide chemical energy for the synthesis of a wide range of complex molecules, including polysaccharides, proteins and oils. Some is stored as starch in specialised tissues and much is incorporated into cellulose, a long-chain condensation product of D-glucose, which is used to build cell structure. Laid down as a mesh of fibrils, cellulose forms the elastic envelope which is the basis of the green plant cell wall. Stretched, thickened, modified by the accumulation of layers of other carbohydrates, notably lignin (a complex, cross-linked condensation product, of great chemical stability), the cellulose walls become the basis of fibre, wood and storage tissues.

Textile fibres

The single-celled seed hairs of the cotton plant are the raw material of one of the oldest, most common and most versatile of textiles, and the purest form of cellulose in everyday use. Multi-celled stem fibres from a variety of plants are the basis of linen, ramie, jute, hemp and sisal. Textiles based on plant fibres are used in clothing, soft furnishings, interior finishes, floor coverings, cords and ropes, padding, packaging and cleaning materials. Individual textile products are relatively homogeneous in structure, with combustion properties influenced by such features as thickness and closeness of weave and texture of surface finish. The latter characteristic is mainly responsible for the rate at which flame spreads over the surface of textiles. Very thin, loosely woven cottons, or knitted cotton or ramie, tend to have loose fibres, which stand out from the general surface. This promotes a very rapid surface spread of flame or 'surface flash' affect when the fabric is exposed to a source of ignition. Fabrics with a raised pile, such as brushed cotton, velvet, corduroy or candlewick, are also subject to rapid surface spread of flame. Standards are now in force in many countries, specifying the minimum fire resistance requirements of fabrics for various end uses [3].

Wood

The woody tissues of tree trunks are built up from bundles of narrow, elongated strengthening fibres and wider water-carrying elements, laid down in annual rings between the bark and the wood. Layered with shorter nutritive ray cells, the whole structure is cemented together with lignin and other complex carbohydrates. The development pattern of wood differs from species to species, with some temperate climate types producing very wide water-carrying elements early in the growing season, followed by much narrower summer and autumn cells. In contrast, tropical woods show little difference in cell diameter throughout the year. Some wood, including both temperate and tropical species, is so coarse-grained as to be almost spongy in texture, while other types are fine-grained and dense. The terms 'hardwood' and 'softwood' tend to be misleading, as the former term is routinely used for the wood of broad-leaved (angiosperm) trees while the latter is applied to conifers, regardless of structure or density. As a tree ages, the inner (older) cells die and empty, to be refilled with tannins and other waste products which darken the wood and increase its density. This 'heartwood' retains much of its density on drying, while the water channels of the younger and paler 'sapwood' are filled with air when dry. Natural wood is therefore heterogeneous in structure, with much variation even in pieces cut from the same tree.

For structural use, wood is cut along the 'grain', that is, parallel to the long axes of the fibres and vessels. Cut into planks, dried, treated with preservative, stained or polished, many different sorts of wood are a principal component of buildings and furniture. Dense, fine-grained conifer woods are mainly used for structural timbers and floors, though oak beams and floors may be encountered in very old buildings and woods such as maple may be used for special purposes. Temperate hardwoods such as oak and beech and the best conifer heartwoods (cedar, pine, etc.) are more likely to be used in good utility furniture and in the decorative finishing of buildings, while tropical and warm-temperate hardwoods are used mainly for fine furniture and veneers (mahogony, rosewood, teak, walnut, etc.).

Wood products

Wood of lesser quality, sliced, chipped, macerated, bound together with synthetic resins, perhaps finished with natural veneers or plastic laminates, is used to produce products such as plywood, blockboard, chipboard, fibreboard and hardboard. These products are more homogeneous than natural wood and have many applications in interior finishing and less expensive furniture ranges for domestic and commercial use. Some high-quality plywoods are used in boat building, roofing and other situations where tensile strength and water resistance are required. Wood-pulp (pine, spruce, etc.) is the raw material of paper, of which vast quantities and many types are in everyday domestic and commercial use.

Other plant products

Stems, roots, leaves, fruits and seeds are all used directly or indirectly for human or animal consumption or as industrial raw materials, and are transported and stored in vast quantities. Plants exploited for human or animal food typically contain large reserves of sugars, starch, oils or proteins and partly processed residues from one industry are often encountered as the raw materials of another.

Animal products

Materials of animal and microbiological origin also abound in daily use. Milk products, animal fats and oils, fermented solids and liquids, dried protein, meat and bone meal, fish by-products, hides and hair, may all be found in large volumes in processing facilities, in storage as large-scale commodities or as contaminants among other commodities. They may be raw or processed, or mixed with raw or processed plant materials as part of manufacture or as contaminants. Wool, a protein fibre, is among the oldest and commonest of textile fibres in general use. Raw wool can be encountered in storage in huge quantities. The processed fibre is used alone or in combination with other animal, plant or manmade fibres in clothing, soft furnishings, carpets and wall coverings. Silk may also be encountered, though in lesser quantities.

Fossil hydrocarbons

In the form of coal, oil, liquefied petroleum gas and natural gas, fossil hydrocarbons are the products of anaerobic decomposition of ancient plant and animal tissues. They are traded and transported worldwide as major commodities for industrial and domestic use. Bulk storage and transport, retail storage and sale, and private storage, of flammable liquids and gases are controlled by legislation in most countries. Legislation and codes of practice in the United Kingdom and Ireland are very similar and closely resemble those in other EU member states and in the United States, Canada, Australia and New Zealand. However, small quantities of both flammable gases and liquids are freely available as solvents and spray propellants and their presence is to be anticipated in all types of industrial and domestic settings.

Natural gas has largely replaced coal gas as a bulk fuel. It is pumped from natural sources and supplied at low pressure to home and industrial users through national and international pipelines. It consists mainly of methane and is lighter than air. Liquefied petroleum gases are a by-product of the petroleum industry and are supplied to the user in pressurised containers of various sizes. Commercial propane, which is a mixture rich in propane but also containing butane and various isomers of both, is supplied in fixed bulk tanks and in portable cylinders in sizes from 11 kg/26 l to 47 kg/108 l. Propane must be stored in the open air. Commercial butane is a mixture consisting mainly of butane, but also containing propane and various isomers of both, supplied in sizes up to 11 kg/26 l and may be stored indoors. It is available in small, pressurised cans for camping stoves, lamps and soldering and other types of gas torches, and in cigarette lighters. It is also used as a propellant in aerosol packages of toiletries, air fresheners, stain removers and paints. The constituents of LPG are heavier than air. Butane in aerosol containers, particularly as a propellant for air fresheners, is sometimes used, mainly by juveniles, as a substance of abuse.

Under legislation in the UK, the Irish Republic and elsewhere, flammable liquids which are stored and distributed in bulk as fuels are classified according to the temperature above which they can form flammable mixtures with air. In the Irish Republic, the relevant classes, defined under the Dangerous Substances Act, 1972 [4] are as follows:

- Petroleum Class I: defined as petroleum which at normal atmospheric pressure gives off a flammable vapour at a temperature of less than 22.8°C.
- Petroleum Class II: defined as petroleum which at normal atmospheric pressure gives off a flammable vapour at a temperature of not less than 22.8°C and not more than 60°C.
- Petroleum Class III: defined as petroleum which at normal atmospheric pressure gives off a flammable vapour at a temperature of not less than 60°C.

Classes I–III are generally equivalent to commercial petrol, kerosene and diesel. Petrol or petroleum Class I is often used as a fire accelerant because of the ease with which it can be ignited. However, this very characteristic makes it extremely dangerous in the hands of inexperienced users (whether innocent or criminal) and its storage and sale, even in small quantities, is controlled by strict legislation in most jurisdictions. Kerosene is freely available both as a bulk fuel and in small quantities. Diesel is mainly available as a bulk fuel.

Apart from their presence in fuels, lower boiling range petroleum fractions such as toluene and xylene have a wide spectrum of uses as solvents, particularly in spray paints, wood finishes, polishes, printing inks, cleaning agents and adhesives. As such, they may be encountered in container sizes from a few ml to 25 l and even 250 l drums in all types of industrial, commercial and residential settings.

Bituminous coal ('smoky' coal) is banned for domestic use in many jurisdictions, because of its contribution to air pollution. Various smokeless fuels are permitted for use in open fires and closed stoves. Smokeless fuels for use in open fires may contain a percentage of petroleum coke for ease of ignition, and may be prone to 'spitting' of hot particles if this is in the form of 'shot coke'. The use of peat as a fuel in the form of machine or hand cut sods or manufactured briquettes is common in Ireland. The dried peat is easily ignited and burns with little smoke. Standards and codes of practice are also set down for storage and delivery of some types of solid fuel and for the performance and installation of appliances for the burning of each type of fuel.

Plastics

In addition to their roles as fuels and solvents, petrochemicals provide feed stock for the manufacture of a vast range of synthetic polymers, the reaction products of naturally occurring carbon-based molecules. They are generally categorised as either thermoplastic, melting on heating, or thermosetting, becoming or remaining solid when heated, characteristics which have a major effect on their behaviour in fires. They are often mixed with inorganic fillers such as glass, asbestos or clays to suit particular end uses. They can also be found in mixtures with organic fibres, as laminates on paper or wood products, in lubricants and as machine parts. Very many have become commonplace as fibres (saturated polyester, nylon 66), packaging film (polyethylene, polyester), storage materials (medium density polyethylene), equipment panels and decorative finishes (acrylics, rigid polystyrenes, melamine formaldehyde), upholstery foams (polyurethane), building insulation materials (expanded polystyrene, rigid polyurethane), floor and seat coverings, water pipes, electrical insulation (polyvinyl chlorides) and electrical fittings (urea formaldehyde). Phenolic resins have a large range of applications, including mouldings, adhesives, hard finishes and foams. Glass-reinforced unsaturated polyesters are used as translucent panels in building construction and are moulded to form rigid structures such as pleasure-boat hulls, machine housings and building panels. Polystyrene, polyurethane and phenolic foams are used as insulating cores in steel construction panels.

Ignition

Flaming ignition

'Ignition may be defined as that process by which a rapid, exothermic reaction is initiated, which then propagates and causes the material involved to undergo change, producing temperatures greatly in excess of ambient' [2]. If the rate of energy release is high enough, flaming ignition will occur. Ignition of gases and volatile liquids always results in flaming combustion. Towards the lower limits of flammability (see Table 2.1) excess of oxygen may result in a very rapid release of energy, with rapid temperature rise and expansion of combustion products. If this takes place in a confined space, a gas or vapour explosion may occur. Towards the higher limits of flammability, flaming is accompanied by smoke with much unburned cabonaceous material in the form of soot particles.

Gases

A gas, by definition, exists in the vapour phase at room temperature. Ignition of flammable gases will occur when a gas/air mixture within the limits of flammability is exposed to a level of energy sufficient to initiate the oxidation reaction. The energy can be supplied by a flame, a mechanical or electrical spark, or a heat flux from a hot surface. Thereafter, the reaction will produce sufficient energy to be self-sustaining, as long as the proportion of gas to air remains within the limits of flammability.

The development of flammable gas as a fuel depended on the availability of technology to contain the gas safely and to control the rate of delivery to a suitable burner. Most accidental explosions and fires involving gas are brought about by carelessness or misuse, only infrequently by failure of properly installed gas handling equipment.

Liquids

Flammable liquids require an initial energy input to produce an air/vapour mixture within the limits of flammability. The flash point is the lowest temperature at which an ignitable air/vapour mixture can exist in equilibrium with a volatile liquid at atmospheric pressure. To maintain flaming thereafter, sufficient energy has to be available to raise the temperature of the air/vapour mixture above the firepoint [2]. Low flash point liquids require special handling because conditions for ignition may exist at or below ordinary ambient temperature.

Burning of less highly flammable liquids is facilitated by absorbing some of the liquid onto a wick of porous material of low heat conductivity. When one end of the wick is dipped into a pool of the flammable liquid and the other end is ignited, liquid is drawn up by capillary action to replace that burned off. This is the principle of the simplest oil lamps, first invented thousands of years ago and widely used even in parts of rural Europe down to the Second World War. The simplest lamps used fish or plant oils in a shell or fired clay container, with wicks of fibre or pith. Tallow and wax candles operate in a similar way, with the run-off of excess wax ensuring that burning remains under control. With enclosed candles, a pool of melted wax can form around the wick and if the heat flux is sufficient, flaming will spread from a too-long wick to the surface of the liquid. It is necessary to keep the wick trimmed to avoid this. The use of candles for decorative effect in the home has become widespread in recent years, particularly among younger people.

Use of the more volatile oils produced by the petrochemical industry depended on the development of the necessary technology to control the materials. Kerosene lamps with enclosed fuel reservoirs and glass-covered flames, kerosene pressurised torches ('blow lamps'), portable kerosene burners and space heaters, were commonplace from the mid-1800s until the 1960s, since when they have been replaced by the development of liquefied petroleum gas technology. Modern kerosene or gas–oil burners used for space heating employ an electrical fuel vaporisation or atomisation and ignition system.

Solids – general

The release of volatile substances from the surface of a solid usually requires chemical decomposition. Ignition of solids, therefore, requires a greater initial energy input than ignition of gases or liquids. Combustible materials tend to be of low thermal conductivity (cellulosic materials, plastics), so that heat is not rapidly dissipated from the point of application and the temperature can rise rapidly to ignition level. In order for burning to continue, the surrounding material must be heated quickly enough to provide a continuous stream of volatile pyrolysis products. The time taken for ignition will depend on the mass and physical structure of the solid, the rate and period of heating and the rate at which heat is lost from the surface. In general, thin layers, loosely woven or loosely packed, with upstanding fibres or ridges on the surface, are more easily ignited by a short exposure to a high energy source such as a flame or spark, than dense, closely packed, tightly woven, smooth surfaced blocks of material of the same chemical composition.

Cellulosic materials

Table 2.2 outlines four stages which have been described in the decomposition of wood on exposure to heat [1].

There is evidence that solid wood and wood products such as fibreboard can be pyrolised by prolonged contact with relatively moderate heat sources, such as hot water pipes. It is suspected that prolonged heating, even at the temperature of boiling water, can lead to the formation of charcoal, which in turn is capable of spontaneous heating. It is suggested that 100°C is the highest temperature to which wood can be continuously exposed without risk of ignition and it is advised that only hot water pipes and radiators fired by automatic boilers which cannot be set to permit water temperature above (150°F) 66°C can be installed adjacent to combustible materials without safety clearance [2]. It is generally accepted that 70°C is the temperature above which chemical oxidation of cellulosic materials begins to take over from microbiological and other external sources of heating in spontaneous ignition sequences [3]. The subject of spontaneous ignition is dealt with separately under 'Sources of ignition of combustible solids'.

Plastics

Plastics constitute a large part of the fire load in homes, work-places and public buildings and their ignition also depends on the production of flammable pyrolysis products. The

Table 2.2 Stages in the decomposition of wood on exposure to heat

Temperature	Reaction
< 200°C	Production of water vapour, carbon dioxide, formic and acetic acids – all non-combustible gases.
200–280°C	Less water and some carbon monoxide – still primary endothermic reaction.
280–500°C	Exothermic reaction with flammable vapours and particulates. Some secondary reaction from charcoal formed.
Over 500°C	Residue principally charcoal with notable catalytic action.

behaviour of a plastic in a fire may depend on whether it is thermoplastic or thermosetting. Thermoplastics tend to melt away from a source of ignition, but when ignited may spread a fire by shedding burning drops. The widely used polystyrene foams are thermoplastic and relatively difficult to ignite, as the polymer shrinks away from the heat source. However, once ignition is established, there is rapid surface spread of flame with dense smoke production. Polyethylene melts away from flame, but when ignited it tends to burn steadily with a low, relatively smokeless flame and sheds burning droplets. Thermosetting plastics produce a rigid char, which may be capable of smouldering either on its own or in contact with another material which provides pilot ignition. In some cases surface char may form an insulating layer, which has a fire retardant effect. Some plastics are self-extinguishing unless continuously exposed to an external heat source. In the case of PVC, this must be at a temperature greater than 470 °C. In general, many plastics ignite at temperatures between 400°C and 500°C.

Textiles

Modern textiles may incorporate both natural and man-made fibres, sometimes three or more in a single blend. Fabrics may be layered or trimmed with other materials, including plastics, in garments or drapes. They may be used as covers for assemblies of several materials in upholstery. They will normally be subjected to detergent and water or solvent-based cleaning during a long or short period of use. This may alter the physical properties of the fabric and diminish the effect of flame-retardant treatments.

Table 2.3, adapted from PD 2777 : 1994 [3] describes burning characteristics of a range of textile materials when tested singly under BS 5438 : 1989 [6], which measures rate and extent of flame spread. Tests of cover and filling assemblies used in upholstered furniture and bedding are carried out under several separate British and international standards. Research is ongoing to improve means of testing the flammability of textiles to incorporate end use conditions and to agree standards across international boundaries.

Smouldering

Cellulosic materials and some thermosetting plastics are able to form a porous carbonaceous char when heated, where volatile combustion products are produced but are not ignited. The char undergoes smouldering, a self-sustaining non-flaming combustion, and the volatile products condense as tars on cool surfaces. This type of combustion may occur where there is inadequate oxygen to promote flaming ignition of the volatile products and is often an important phase in the development of accidental fires. Drysdale (1985) [2] has described the smouldering process in a horizontal cellulose rod as follows:

- Zone 1 – pyrolysis zone where there is steep temperature rise and an outflow of combustion products (tars, volatile liquids, smoke) from the parent material. In flaming combustion these products burn as the flame.
- Zone 2 – charred zone where temperature reaches a maximum, typically 600 – 750°C, smoke evolution stops and glowing occurs. Oxygen diffusion determines the rate of heat release.
- Zone 3 – residual ash/char, where temperature is falling slowly.

Table 2.3 Burning characteristics of different textiles

Flammability class	Features	Typical materials	Comments
Highly flammable	Very rapid flame spread with base material burning	Very lightweight or highly raised cotton or viscose	Not allowed as apparel fabrics under US Flammable Fabrics Act
Surface flash	Rapid flame spread over fabric surface, but without ignition of base material	Brushed, raised and knitted fabrics of cotton, ramie and viscose	Surface flash occurs more readily if fabric surface is brushed up and if the fabric is vary dry
Flammable	Easily ignited, burns rapidly and completely	Acetate, acrylic, cotton, linen, modal, ramie, triacetate, viscose	In widespread and common use as apparel fabrics
Low flammability Thermoplastic	Melts on application of heat, may ignite, may give molten drops which may be flaming	Polyester, polyamide, PVC	Burns more readily when supported, e.g. in blends with non-thermoplastic fibres, and when treated with certain prints and finishes
Limited flame spread	May ignite, but no flame spread. Forms solid char with no melting	Flame retardant cotton, viscose and wool, modacrylic	Properties may be inherent or achieved by durable or non-durable flame retardant treatment
Not ignited	Does not ignite	Aramid, Polybenzamidizole (PBI), carbon and glass fibres	Used in heat protective equipment

This process can be observed in the burning of a cigarette. In practice, in enclosed spaces, dry wood and other plant materials in which fibrous tissues are preserved, are capable of smouldering indefinitely, for as long as fuel is available and external factors tend to conserve heat. The internal air supply in dried plant tissues undoubtedly supports combustion in situations where it would otherwise fail for lack of oxygen. Personal observation, reinforced by experiment, has identified the following as materials in which smouldering ignition is both easily established and persistent: jute, whether as loose fibres or woven into sacking; cotton, particularly in the form of soft-woven fabrics; soft paper such as kitchen paper, facial/handkerchief tissues or toilet tissue; corrugated cardboard packaging; wood shavings, whether loose as manufacturing waste, as packing, or as filling or stuffing in older types of upholstery; household 'fluff', which is a collection of many types of plant, animal, micro-biological and inorganic debris; sweepings or the contents of vacuum cleaners, which may contain fragments of any or all of the above. These materials provide a tinder in which a short-lived ignition source such as a spark from metal cutting or welding, a small piece of cigarette char or a brief flame contact, can initiate smouldering combustion, capable of developing to involve larger fuel masses.

The relatively low ignition temperature of lightweight cellulosic materials means that hot combustion gases can ignite structures such as cotton net curtains and paper lampshades at some distance from an original fire origin, creating a false impression of multiple fire seats. The metal hooks/wires which support the textiles normally survive.

Sources of ignition of combustible solids

Smokers materials

A common cause of fire in both domestic and industrial settings is lack of care in the handling and disposal of smokers' materials. Cigarettes and matches are ready sources of ignition, each producing characteristic fire development sequences. The forgotten cigarette falling off an ashtray when its centre of gravity changes and the fire caused by emptying an ashtray onto combustible materials in a waste basket are images used to promote consciousness of fire safety, but such incidents are still all too frequent. Where a match or cigarette char is dropped onto clothing or bedding, the occurrence can have very serious consequences for the smoker. Such accidents may result in serious injury or death, particularly if the victims are children or elderly people, perhaps slow to recognise the danger or unable to summon help quickly. Research in the United Kingdom has identified older people as particularly at risk from fires involving clothing and bedding [3]. Younger age groups are also at risk.

A personal friend who liked to smoke while reading in bed, fell asleep with a lighted cigarette in his hand and awoke to find his bedding smouldering. In another case, I was asked to examine a garment which had been ignited by a match flame. The wearer, a young woman, had been attempting to light a cigarette while driving, and had dropped the lighted match onto the floor of the car. It had ignited her lightweight cotton dress and she suffered serious burns before managing to bring the vehicle to a halt and escape. In another case, a 3-year-old girl, left in the family car while her mother unlocked the house door, managed to ignite the fluffy polyester seat cover with a match from a box left under the dashboard. The mother returned to find the vehicle full of smoke and the child with injuries, which eventually resulted in loss of parts of the fingers of both hands and severe facial scarring. Her woollen coat and tights saved the rest of her body from serious injury.

In personal experiments, it was found that freshly lighted filter-tipped cigarettes of standard length would smoulder unattended for between 10 and 20 min, with exposure to air currents tending to accelerate the rate of burning. In a test carried out as part of the investigation into a warehouse fire, a half-smoked cigarette was placed on top of a bale of toilet rolls packed in polyethylene film. The lighted end penetrated the outer and inner polyethylene wraps immediately and the paper began to smoulder as soon as the hot char touched it. In a light air current, the paper smouldered for about 10 min, filling the top of the pack with smoke and producing a red glow. After 12 min, the toilet rolls burst into flame and after 14 min the fire was out of easy control. In the test, a hose was used at this stage to douse the flames, but the fire re-ignited several times as soon as the water was turned off.

The following incident occurred while a party of ladies were awaiting afternoon tea in a hotel lounge. Two were seated facing one another, in high-backed chairs with an additional loose cushion at the occupant's back. A third stood chatting to them, leaning on the back of one of the chairs, lighted cigarette in hand. When the tea was served, the smoker moved to join another group and one of the seated pair poured the tea. As she settled back with her own cup, she noticed what at first appeared to be a column of steam rising past her friend's head. As the column increased in volume, she realised that she was observing smoke, and that it was rising from behind the other woman's back. She alerted her friend, who sprang from the chair and pulled out the cushion. A hole roughly 100 mm in diameter and 20 mm in depth had been burned where the loose cushion met the padded chair-back. The fire was extinguished with the remaining tea and a jug of water brought by the waiter. The time between the departure of the smoker and first sighting of the smoke column was estimated at about 5 min. The chairs were foam-upholstered, with cotton covers, and pre-dated modern fire-resistance standards. The women agreed that the fire must have been caused by a particle of hot char from the cigarette, falling between the chair-back and the loose cushion. However, none of the three involved had been aware of any fall of char at the time.

The next case is one of many in which hot char was included in waste, where it started a smouldering fire which later spread to other materials. In this instance, what was described as fresh ash was gathered up with a vacuum cleaner. Miss N., in her twenties, lived with her parents and lunched at home every day. Mr N. was very much opposed to smoking and during the investigation expressed his total opposition to smoking in the house. However, Miss N. usually smoked a cigarette at the end of her lunch and in order to preserve family harmony, her mother vacuumed up any evidence in the form of dropped cigarette ash. Mrs N. insisted that she would not suck up a cigarette end, but would clear up particles immediately her daughter dropped them. Afterwards, the vacuum cleaner was put away in a hallway leading to the back door, in an area also used to store newspapers and hang coats. One evening, as the family were watching the news on television, they became aware of unusual sounds in the rear hall and discovered that a fire was in progress. The Fire Brigade were quickly on the scene and brought the outbreak under control. Investigation of the debris indicated that the fire had spread from floor level beside the vacuum cleaner, which was found partially intact under fallen coats. The plastic body of the cleaner was penetrated near the bottom on one side, and the rest had softened and folded over the paper dust bag. When it was cut open, it was found that the cleaner body was smoke-stained on the inside and that there was a pocket of charred dust inside the dust bag, which was still about two-thirds full. It appeared that about 5 hours after the cigarette ash had been gathered up, smouldering dust had melted through the lower part of the plastic body of the cleaner and had come into contact with the newspapers. The resulting fire had spread to the coats hanging above, some of which had fallen off their melting plastic hangers, alerting the family and preserving the evidence in the cleaner.

Smoking is prohibited in many workplaces, but breakdown of smoking discipline is often associated with departures from the normal work routine. Any unusual work situation such as overtime, particularly at weekends and if the worker is alone, stocktaking, maintenance work during holiday periods, can lead to a breakdown in normal discipline in relation to smoking regulations. The next two cases illustrate this problem, which has been found to apply at all levels, from shop floor to senior management and boardroom. It should be addressed directly as part of fire safety training.

A four-storey building housing an aircraft maintenance business was the scene of a potentially disastrous fire in the early 1990s. Most of the building consisted of a large aircraft maintenance hangar, with a four-storey stores and administration block to the front. There were two large passenger aircraft in the hangar at the time of the incident.

Stairs and lifts at the front of the building gave access to the upper floors, where a suite of communal and individual offices occupied the top floor frontage. The floors were of concrete slab construction, and the office floor was covered with industrial quality carpet. There was a suspended ceiling of standard light steel and mineral slab construction and a row of windows along the external wall. The offices were connected, without fire stops, by an open circulation area along the inner wall. There was a single exit to the lifts and stairwell.

The communal offices were laid out in a number of workstations, grouped in units of four. Each station consisted of an L-shaped desk, with raised back and sides separating it from the stations beside and opposite it. The desk and its ancillary units were constructed of laminated and PVC-covered chipboard on steel frames. At one end, there was a shelving unit above the desktop and a drawer unit below it. There was a wastebasket under the other end of the desk. A personal computer stood in the apex of the L-shape and there were also telephone facilities on the desk. Each workstation was provided with a standard revolving office chair, upholstered in PVC and polyurethane foam on a steel frame. The wastebasket was of polypropylene and there were various other plastic materials in the PC and telephone equipment. Electrical and electronic services were provided in PVC trunking along the base of the outer wall of the office compartment. There were also cable runs above the ceiling, serving security systems (including smoke detectors) and lights.

A female employee was working alone in the office furthest from the stairwell on a Sunday morning. She left her desk at about 11.00 a.m. to go to the canteen. There was a smoke detector above the workstation next to that at which the woman was working. The fire alarm sounded at 11.07 a.m. Coincidentally, an aircraft towing truck in the maintenance hanger suffered an engine misfire, which caused it to emit a cloud of exhaust smoke. It was thought that this had activated the alarm and so the alarm was silenced. The alarm activated again a short time later and was again silenced. At 11.17 a.m., a passer by reported to the reception desk that smoke was issuing from the top floor windows at the front of the building and the emergency services were finally alerted.

Inspection of the scene indicated that the fire started at the workstation at which the staff-member had been working shortly before the first alarm. Examination of the PC and the incoming electrical services uncovered no evidence of malfunction. They had been burned from the outside, as had the conduit in which the electrical wiring was laid. Much of the debris had been removed by the Fire Brigade, but marks on the carpet indicated where the bases of the drawer unit and bin had remained intact. It was clear that the desktop, shelving unit and most of the drawer-unit had been burned away by a fire coming from the direction of the waste bin. The fire had spread into the adjoining workstations and destroyed part of the ceiling immediately over the fire origin, but otherwise damage to the fabric of the building was relatively minor. However, there was widespread and serious smoke logging throughout the entire office floor, resulting in disruption of computer function throughout the offices.

When presented with the evidence, the woman admitted that she had smoked a cigarette at her workstation, although smoking was banned in the area and no services for smoking were provided. She had stubbed out the cigarette on a piece of paper, which she crumpled up with the butt inside and dropped into the waste basket at her desk. She then left the office to go to the canteen. Although failure of staff discipline was responsible for the initiation of this fire, building layout and materials and failure of fire safety procedures contributed to the extent of the loss.

The second incident occurred while stocktaking was in progress in premises consisting of a large, glass-fronted, car showroom and workshop, with associated stores and offices. The offices, constructed of prefabricated chipboard and reinforced glass panels, with chipboard floors and ceilings, were built against the block wall separating the showroom from the workshop. The walls were painted and the floors were covered with carpet tiles. The dealership building, including showroom, workshop and stores, was closed at weekends, while the forecourt remained open for fuel sales. There was a strict ban on smoking in the work areas.

The stocktaking operation had begun with a Friday consultation attended by the firm's auditor. Two members of the firm's middle management worked overtime with the clerk-receptionist on the Saturday, to produce documentation for presentation to a further meeting scheduled for the following Monday. The parts manager and the chief salesman spent the day moving between the clerk-receptionist's office and the stores, checking the stock and preparing lists, while the clerk-receptionist compiled the stock reports on her computer terminal and made photocopies. The computer desk was located in the corner between the doors to the stores and the showroom, with the photocopier on a cabinet against the block wall. In the clerk-receptionist's words, by the end of the day there were piles of reports, invoices and other documents all over her office, including on the floor.

The clerk/receptionist was the last person to leave the building. She had switched off all the equipment and the lights, locked her office and the external doors and left the building at about 17.00 h. The fire was discovered shortly afterwards by a forecourt customer, who glanced through the glass panels at the front of the building, saw flames in the office and raised the alarm. The Fire Brigade recorded a call at 17.10 h and the first tender was on site 8 min later. The brigade found the doors locked and all external glass panels intact.

This fire was detected at a very early stage, and the Fire Brigade were on the scene within 8 min of the alarm being raised. Nevertheless, the outer partition walls and the contents of the office were severely damaged. The combustible parts of the furniture were burned out, papers which had been stacked on the floor, chairs and desks were completely charred, the partition was burned to floor level and the floor was penetrated in the corner beside the computer desk. There was also damage to the stores and to two of the new vehicles in the showroom. The burning patterns indicated a fire origin at or close to floor level in the area in the corner between the doors to the stores and the showroom, where the staff were working during the day. On examination, it became clear that an electrical cause could be excluded, leaving the question of staff involvement.

When the possibility of smoking was raised, the Manager was adamant that the strict ban on smoking in the work areas was respected by all his staff. However, the staff members involved freely admitted to smoking during the stocktaking, because it was not regarded as normal work. Smoking had started on the previous day, when the auditor had asked for an ashtray during the preliminary meeting. The staff involved in the stocktaking had also lit cigarettes and all three continued to smoke on the following day. The clerk-receptionist admitted that she had smoked throughout the previous day and on the day of the fire, although she did not normally smoke at work. She could not remember whether the ashtray had still been in her office on the day of the fire and did not remember using it on that day. She had stubbed out a cigarette on the carpeted floor just before leaving the office by the door to the showroom, as she did not wish to carry it through to the forecourt for safety reasons. The fire was discovered 10 min after she had left the premises. Even where occupants do not admit to smoking, or to carelessness with burning cigarettes, it is possible to find evidence such as dropped cigarette ends and spent matches in surviving parts of the premises.

Matches

The most common use of matches is to light cigarettes, cigars and pipes and in this context, lighted matches are freely available as a means of ignition. Many users blow out a match flame with their breath or by shaking the match vigorously, before dropping it in a bin, or placing it in an ashtray. Some drop the match on the ground, with or without extinguishing it first, particularly if they are out of doors or in a large space such as a store or factory building. From time to time the flame will flare up again, perhaps when the match has been dropped on the ground or into a bin. If it has fallen onto flammable materials, fire may develop rapidly. Waste paper or other refuse may be ignited, spreading the fire to other fuels. In general, fire started by a match flame will result in an immediate outbreak.

In some cases, matches may be lighted and discarded in areas where naked flame is not or should not be permitted. The late Dr Diarmuid Mac Daeid and I carried out *ad hoc* tests to determine (a) how far a match was likely to travel if thrown casually over the shoulder; (b) if discarded while still burning, how likely was it that a match would remain alight on falling to the ground; (c) if a lighted match were discarded among paper strips, how likely was it to cause ignition (the choice of materials was determined by the case under investigation at the time). The tests were carried out inside a building and in tests (a) and (b) each participant threw two batches of 100 matches. We obtained the following results:

- Matches thrown over the shoulder landed in a rough circle, centred at a distance approximately equal to the height of the thrower. About 70% of the matches landed in a circle about 1 m in diameter and all were contained in a circle whose diameter was approximately equal to the height of the thrower.
- Of 100 lighted matches thrown over the shoulder, an average of 3 continued to burn on landing. (It was considered unsafe to carry out this test with papers in the target area.)
- Of 100 lighted matches dropped vertically onto paper strips from a height of 750 mm, 20 caused ignition.
- Of 100 lighted matches thrown onto paper strips from a distance of 1 m, 11 caused ignition.
- Of 10 lighted matches thrown a distance of 3 m, all extinguished in the air.

This exercise demonstrated that a fire caused by a discarded lighted match usually occurs close to the person responsible. Few fires accidentally started by lighted matches go unnoticed for long enough to cause a serious outbreak, unless they occur in areas where naked flames should not be permitted. The only site I have visited where those entering were required to surrender matches, cigarette lighters and camera flash bulbs before passing the security barrier was a petroleum fuel depot. On that site there was a clear possibility that dangerous concentrations of flammable vapours could be present, but there are many sites where failure to enforce absolute prohibition of naked flames in the workplace has resulted in unnecessary and disastrous fire development.

A serious fire occurred in a factory making expanded polystyrene blocks for packaging and insulation slabs. It resulted from a practice which had grown up, of allowing staff to take smoking breaks at the goods-inwards ramp. Workers walked to the open door through a storage area where large freshly manufactured blocks of polystyrene foam were stood on end to cool before going to the cutting area. According to statements made afterwards, the area was kept clear of general rubbish, but small quantities of loose polystyrene beads tended to collect around the bases of the slabs. A few minutes after a worker had passed through on his way for a smoke, a fire was noticed among the blocks nearest the door to the ramp. The worker later claimed to have first seen the fire in a different part of the floor and to have raised the alarm and attempted to deal with it. This was at variance with independent statements from the two others working in the area. The outbreak spread so rapidly that the rest of the staff had to flee for their lives and the entire building and contents were destroyed. There was no electrical involvement and no cause could be identified other than a carelessly dropped match or deliberate sabotage. There appeared to be no reason in this case to suspect the latter. The worker freely admitted that he had walked through the area to have a smoke, but denied dropping a lighted match inside the store. In this case, there was a failure on the part of management as well as workers, in allowing dangerous practices to become the norm.

In a similar case, a small fire which started in packaging waste spread to engulf most of a meat-processing plant. Again, smoking was at the root of the matter and a junior worker admitted that he had been smoking nearby and had thrown away a lighted match. However, he claimed that he had thrown the match away while standing outside with his back to the door, a distance of about 5 m from the fire origin. He claimed that he saw the fire, but did not report it immediately, fearing retribution. Instead, he closed the door on the blaze and claimed he thought it would die away. By the time older colleagues noticed smoke and went to investigate, the fire was out of control.

In both the cases described here, smoking discipline had been allowed to lapse, so that from a position where workers were allowed to smoke just beyond the margins of the work area, they had begun to light cigarettes as they went towards the smoking zone, and this behaviour was tolerated by management. Both fires resulted in property losses running into millions of pounds, interruption or loss of employment and environmental pollution.

Solid fuel

Open fireplaces and traditional solid fuel stoves add charm and comfort to a room. They can present particular dangers however, due to the necessity to add fuel and handle hot ash and cinders manually. Building regulations in most countries prescribe standards of construction designed to prevent heat transfer from the hearth or stove to underlying or surrounding combustible materials. In addition, a fender or similar protective device should be used to contain any hot fuel which escapes from the grate or fire box. Open fires must not be left unattended, unless provided with a mesh guard fine enough to prevent the escape of sparks from whichever type of solid fuel is used. Some types of wood and solid fuel eject hot particles while burning; these can be thrown onto nearby carpets, furniture or the clothing of persons standing or sitting nearby. If the household includes elderly people, children or domestic animals, special care is necessary and a fireguard which can be secured to the wall on both sides of the fireplace and which is fitted a strong, secure top, is an essential safeguard. The following incident involved an open fire, which was inadequately guarded.

Shortly before going out to spend an evening with her sister, Mrs C. lit a fire of logs and solid fuel nuggets in the open grate in her lounge, so that the room would be warm when she returned. The grate was the type with bars at the front, insufficient to prevent hot embers of the smokeless fuel from falling through. There was neither fender nor guard in front of the fireplace. The lounge floor was 3/4 inch tongue and groove boards, covered with a wool carpet and underlay. The furniture included a lounge suite, a table-lamp and a CD player, the latter standing on the carpet. Both appliances were connected to the electricity supply and the CD player was on standby.

Mrs C. was absent for about 5 hours, and returned to discover that the floor of the lounge had been almost entirely burned away. The burning appeared to have spread under the floor, with little spread of the fire out of the compartment, and fittings at the perimeter of the room surviving almost intact. The plaster-slab ceiling had been penetrated above the area where the fireplace and also the lamp and CD player were located. Although the joists and timbers of the upper floor were heavily charred, the outbreak was discovered before it had broken through to the upper storey. The fire did not appear to have been electrical in origin. There was relatively little wiring under the floor, and there was no evidence of electrical fault on any part of it. The wiring at the back of the sockets was intact. The remains of the CD player were recovered from the debris and the appliance was clearly burned from the outside. The intact base was resting on a piece of carpet which had been charred from below and which was resting on charred fragments of the floor.

There was no evidence to indicate that the fire was other than accidental in origin. It had smouldered probably without developing to flaming ignition at any stage. The most likely cause of the outbreak was ignition of the carpet and floorboards, by embers of smokeless fuel or a log falling from the unguarded open grate. The fire burned through the floor and then continued to smoulder under the floorboards. The smokeless fuel was sold in bags, which carried a warning indicating that a spark guard should be placed in front of open fires.

The susceptibility of lightweight textiles to ignition in the vicinity of open fires is illustrated by the following two cases involving children. In the first incident, a 3-year-old girl sustained severe injuries when her clothing caught fire while she was visiting her grandparents' home. The child walked close to an unguarded open fire while wearing a lightweight polyester and cotton coat. The back of the garment was ignited, and although her grandfather quickly extinguished the flames, the child received severe burns to her upper legs and back. The accident occurred in the Irish Republic, while the garment involved was manufactured in the United Kingdom for an Irish retail outlet. A duplicate garment produced by the retailer failed to meet the requirements of the nightwear regulations current in both countries when tested by the methods described in BS 5438 : 1989 [5]. In this case, the Irish Courts held that the grandfather was negligent in failing to provide a guard for the fire, but that the manufacturer and retailer also had a duty of care. It was held that although there were no regulatory performance requirements for children's day wear, failure to meet the nightwear standard indicated that the article was highly flammable and as a matter of prudence, should have carried a fire warning label. In another case, a 10-year-old boy was at home alone while his mother had gone a few doors away in response to a call from her elderly mother. The boy was playing with a video game while sitting on the floor, in the centre of an average-sized living room, with his back to an unguarded open fire of smokeless fuel. The fuel was supplied in bags printed with a warning that a spark guard should be used. He was wearing an official merchandise replica Premier League football shirt, produced in the UK by a leading sportswear manufacturer. The shirt was made from 100% polyester knitted fabric and carried fire warning labelling similar to that required under the nightwear regulations. Nevertheless, four standard samples from a duplicate garment, tested by the methods described in BS 5438:1989 [5], met the performance requirements of the nightwear regulations (not enough fabric was available to carry out a definitive test). A piece of burning fuel burst, throwing hot embers across the room. An ember stuck to the child's shirt, causing the fabric to ignite. Although he managed to pull the shirt off and was rendered prompt and correct first aid by a neighbour who heard the house smoke alarm, the victim suffered very severe burns to his back and arms, necessitating a week's treatment in intensive care and a total of four-and-a-half month's hospitalisation. This case illustrates that the provision of safety warnings by manufacturers and retailers is futile, unless the public are educated to read and take note of product labelling.

The involvement of smokeless fuel in two of the last three examples was to some extent coincidental, as the damage and injuries sustained depended on the propagation of flame or smouldering in the burning material following contact ignition. However, following the phasing out of the use of bituminous coal in the Dublin area in the early 1990s, the use in open fires of smokeless fuel incorporating a percentage of petroleum coke led to a number of fires in older houses. The available smokeless fuels for use in domestic open fires have a higher calorific value than coal. They burn at a higher temperature and produce more radiant heat, even though they may meet the Irish legislation as set out in Industrial Research and Standards (Section 44) (Petroleum Coke and other Solid Fuels) Order, 1991. As a result, hearths in older houses, designed to withstand the lower temperatures of coal fires, have in some instances been cracked, while in others it appeared that the radiant heat had penetrated through thin hearths or fire-backs and gradually dried and pyrolysed timbers installed under or behind them. Twelve fire outbreaks associated with the use of these products, in houses dating from the mid- to late-eighteenth century, were reported to the Institiute for Industrial Research and Standards (personal communication) in the two years following their introduction. The following account is typical of these incidents.

The fire occurred in one of a terrace of late eighteenth-century houses standing four stories over basement. The building and its neighbours had been redeveloped in the early twentieth century into a number of apartments, including single bedsits. The fire had affected two such apartments, one above the other. The original fireplaces were still in use in the rooms affected by the fire. They were small, with cast-iron chimney pieces which incorporated what appeared to be the original cast-iron and fire-clay backs. The hearths consisted of a concrete slab reinforced with slates and between two-and-a-half and three inches thick. The inner edge was supported in the wall below the chimney and extended at the front into a box formed by timbers mounted at the front in the floor joists and at the sides in the wall masonry on either side of hearth. In front of the fireplace, the top of the concrete hearth slab was finished with tiles or slate to bring it level with the floorboards. The occupant of the upper flat used the fireplace regularly as the main source of heating. In the early hours of a December morning, the occupant of the lower flat became aware of smoke in his room and raised the alarm. The Fire Brigade discovered a fire in progress in the void between the floor of the upper flat and the ceiling of the room below.

Examination of the electrical wiring in the floor void indicated that it had been damaged by the fire and was not responsible for the outbreak. The ceiling light fitting in the lower room was clearly burned from outside, the copper conductors remaining intact although the insulation had been burned off them where they crossed the fire zone. Within the light fitting the terminals remained intact and the wiring insulation unburned.

The fire had originated in the timbers enclosing the hearth in the upper room. The joist and the hearth timber running into the masonry at the right-hand side of the fireplace were charred away and the fire had spread along the front of the box supporting the hearth to involve a second joist. Burning continued to spread outwards under the floor along the two joists so that the joist immediately to the right of the fireplace was consumed to a point near the centre of the floor. The second joist was charred for a shorter distance, and eventually part of it fell out of its socket. The fire was contained by the ceiling of the lower room, which was formed of two slabs of plasterboard with a plaster skim, rated to give a half-hour fire resistance. It remained intact and helped to reduce the air supply to the fire even when the supporting timbers had burned through. A heavy carpet on the floor above also helped to make the floor cavity airtight and slow the fire-spread. This confined the outbreak to the vicinity of the fireplace until it was extinguished. The occupants of the two apartments were unaware of the developing fire, because at first any smoke produced was drawn into the flue. Eventually, the fire broke through the floor beside the upper fireplace, probably when charred timbers collapsed inside the floor cavity and allowed air to penetrate at the side of the hearth. A plume of soot on the wall beside the fireplace in the upper room was the only major smoke stain in either room. Most of the debris remained in place until pulled down by the Fire Brigade.

Candles

The obvious danger of naked flame associated with candles is generally thought to be less if the candle is enclosed in a metal cage, and possibly covered with a glass or even plastic screen. The original 'nightlight', consisting of a small flat candle in a light steel cup, was intended for nursery or sickroom use. It was placed in a saucer of water and might be used uncovered or with a wide glass guard. The same type was used, set in a small glass bowl, as a votive lamp at religious shrines. The new vogue for use of candles as a fashion item has seen a great increase in their use for home decoration. However, many users have very little awareness of the potential of apparently 'safe' candles to cause fires. The nightlight or tealight consists of a metal cup filled with wax into which a wick is set, anchored in a steel strip set in the bottom of the wax. The outer cup gets hot as the candle burns. The manufacturers' instructions issued with such candles include a warning not to place the night light on a surface which can be damaged by heat, but to set it on a heat-proof base, preferably in a saucer of water, before leaving it unattended.

In one case where a nightlight was suspected of causing a fire, it transpired that the candle was not set on a heat-proof base, but was placed on the unprotected plastic outer casing of a television set, with a polyethylene Christmas ornament placed over it. The ornament was made in the shape of a Christmas tree, with small holes representing lights. It was designed to fit over a nightlight, with the candle light showing through the holes. The fact that the nightlight was covered by the Christmas ornament would have helped it to retain heat, but otherwise probably made no real difference to the outcome. The family left the house, leaving the candle burning. The damage was caused by the hot base of the nightlight, which melted through the top of the television set and allowed the burning candle to fall through into the interior of the set, where it ignited combustible plastic materials.

In a second case, a new television set with a plastic outer case was placed on top of a wooden cabinet in the living room. At the time of the fire, it was connected to the house electricity supply by a plug-top containing a 5 A fuse, but was switched off. At about 03.30, one of the children awoke to find that her room was full of smoke and she experienced some difficulty in breathing. She also heard the smoke alarm sounding and immediately aroused the rest of the family. All left the building safely, although there was thick smoke in the house at that stage. The entire interior of the house was heavily smoke-blackened.

Examination of the television set indicated that although many of the components were cracked and most were embedded in melted plastic from the case, they remained sufficiently intact to indicate that the fire had started outside the set and had burned downwards into the back of the casing. The tube had shattered and some sections of aluminium at the back of the set had melted. Some of the components were entirely intact, though covered in soot and melted plastic from the casing, but some of the plastic components smouldered, creating the large volumes of smoke which had clearly filled the house. The fire had also slightly damaged a radio cassette player, but had left the cabinet, on which both appliances were standing, practically intact. It emerged that when going to bed on the night of the fire, the family had left a nightlight candle burning on top of the television set. It appeared that the hot metal base of the candle had eventually melted the plastic casing of the television and allowed the nightlight to drop through and burn the interior of the set.

With larger candles, the wick needs to be kept trimmed to avoid a dangerous increase in size of the flame. If left untrimmed, the wick can arch over and touch the melted wax, so that a double wick affect is created. This can cause the candle to flare up to such an extent that nearby materials may be ignited. A fire started by a candle left alight overnight in a young man's bedroom resulted in his death from smoke inhalation. The burning patterns indicated that the fire started on a table, where he also kept papers and magazines. He and his girlfriend would chat by candlelight on their mobile phones, then leave the candles to burn overnight while they slept, as a romantic gesture. The danger can be greater with some craft candles, which have been decorated on the outside with paper wrappings or cut-outs. Candles of this sort should never be left unattended unless placed in a container of water and kept well clear of combustible materials. Candles and matches should never be left within reach of small children.

Hot work

Under this category are collected industrial operations including mechanical cutting and grinding, welding, and the use of gas torches, propane and butane as well as oxyacetylene. All the operations produce heat sufficient to ignite flammable vapours and to cause flaming or smouldering in susceptible materials. The gas flames of oxyacetylene or oxypropane equipment are sufficiently hot to melt steel and other metals, while those of the propane-fuelled torches used in roofing work reach 1200°C. The 'sparks' thrown out by welding, cutting and grinding equipment are in fact droplets of molten metal, heated to incandescence, capable of igniting flammable vapours and of causing smouldering or even flaming ignition in materials such as soft papers, fabrics and some plastic foams, if they come into contact with them while still sufficiently hot.

The recommendations of the Loss Prevention Council [7] provide a code of best practice drawn up by the insurance industry for these operations in the United Kingdom and the Republic of Ireland. A similar code is provided by the Factory Mutual Engineering Corporation booklet 'A Pocket Guide to Hot Work Loss Prevention' 1997 [8], based on US experience. Both codes detail the dangers inherent in these operations and the precautions necessary if they are to be carried out safely. The following are some of the general recommendations from the LPC code:

1. *Location*
    1.2 The area within 15 m of the process should be free of combustible materials before work commences. Where combustible materials situated within 15 m cannot be removed, they should be protected by non-combustible blankets or screens.
    1.3 Work should never be carried out in an atmosphere containing flammable vapours or combustible dust.
2. *General procedure*
    2.2 The work should be carried out only by, or under the supervision of, trained personnel. At least two persons should be present during such work.

2.3 Combustible floors in the segregated area should be wetted down and liberally covered with sand or protected with overlapping sheets of non-combustible material. Particular care needs to be taken to see that any gaps in flooring are adequately covered so that sparks cannot fall into concealed spaces.

2.4 Before carrying out any work on one side of a wall or partition an inspection should be made of the area on the other side to ensure that any combustible materials present are not in danger of ignition either by direct or conducted heat.

2.7 A thorough examination should be made in the vicinity of the process continuously during a period of 1 hour after the termination of each period of work, including an inspection of the area described in paragraph 2.4.

7. *Fire extinguishment*

7.1 At least one portable fire extinguisher having a minimum rating of 13 A or a hose reel should be provided and made ready for use in the event of outbreak of fire. It is also recommended that wet sacking be kept available for immediate use near the scene of work.

7.2 All fire extinguishing equipment should be maintained in efficient working order and be kept in readily accessible positions.

7.3 An employee trained in the use of fire extinguishing equipment should be present during the work and all personnel should be familiar with the method of raising the fire alarm at the premises.

Both codes advise the use of a 'hot work permit' system to verify that all the recommended precautions have been taken before the work commences. They point out that the same precautions must be taken with all hot work, no matter how small the operation may be. In addition to the hot work recommendations, the operation may be subject to insurance warranties in regard to waste management and storage of flammable liquids. The following incidents illustrate the types of accident which can arise from failure to observe the necessary precautions.

A small furniture firm occupied a unit in a local industrial centre. The company manufactured bar and restaurant furniture, constructing the smaller timber frames, buying in metal and heavier timber frames, machine-finishing both wood and metal and adding materials including paint finishes, decorative panels and foam and fabric upholstery. The principal machine tools were housed in the main joinery workshop, while the fixed machinery in the separate finishing area included a planer, a band-saw and a panel-saw. An angle grinder and a hand router were among a number of portable power tools used in this area, where upholstering, painting and spray painting also took place. Timber was stored on a steel rack to the left inside a sliding door at the front of the finishing area and an enclosed booth for spray painting was to the right. The booth was constructed from sheet steel, with an extractor system venting to the outside of the building. The main compressor was housed separately outside the building. There was a small office at the front of the building.

Materials containing flammable liquids and vapours, including adhesives, lacquers and other wood-finishing products were in a store, well removed from the main building. Small quantities were drawn for use on a daily basis. At the time of the fire, there were five or six part-full 5-litre tins in the spray-booth, as well as some tins of ordinary gloss paint for use on steel frames. The machine tools and the router produced off-cuts, shavings and dust, and waste upholstery foam and fabric also accumulated. It was usual practice to clean up waste into bins from time to time during work periods, but there were small accumulations of freshly produced saw-dust and shavings around the machines at the time of the fire. Full bins were placed outside the building and were emptied at the end of the week.

At mid-morning on the day of the fire, a number of employees were working near the open doorway of the finishing area, because the day was very warm. Two were working beside the spray booth, which was not in use at the time. They were using white spirit from a 2-gallon container to clean tubular chair frames in preparation for re-painting. Another was operating an angle grinder, using it to clear rusted or cracked pieces from another set of chair frames. As is usual, the grinder was throwing out showers of incandescent metal droplet 'sparks'. The foreman witnessed sparks from the grinder igniting a thin layer of saw-dust on the floor and saw the fire spreading in the direction of the workers who were using the white spirit. He shouted to them to leave the building while he himself tried to put out the fire with an extinguisher. He was unable to prevent the fire from spreading to the white spirit and the spray booth. The Fire Brigade were called, but on arrival they had trouble in securing an adequate water supply in this rural area. Some office equipment and files were saved from the office before the fire spread throughout the entire building, destroying the structure and contents.

Examination of the debris in the premises indicated that as claimed, only a thin layer of saw-dust had accumulated in the vicinity of the machinery, but in the warm and dry conditions prevailing on the day in question, this material may well have been in a state which might be described as tinder-dry. White spirit is a flammable liquid, but its flash point of about 38°C means that it is stable and safe to use even in warm ambient conditions. Although its use in the normal course of events would not pose any great hazard, its presence at the scene of the established fire undoubtedly contributed to the spread of the outbreak. The management of the spray lacquers, which contain highly volatile solvents, flammable at ambient temperatures, appeared to have been adequate. The containers in the store remained safe during the fire and the quantity in the spray booth was relatively small. As with the white spirit, the presence of any flammable liquid within the premises must have contributed to the spread of fire, once it became involved.

Although there was a fire extinguisher to hand, it proved inadequate to deal with the fire even in its early stages. Few fire extinguishers are capable of cooling a substantial fire adequately, with the result that re-ignition occurs immediately after the discharge is finished. The purpose of extinguishers is mainly to keep fire at bay long enough to allow people to escape. In commercial premises where timber and other combustible materials provide a large fire load, the provision of a fire hose is necessary to give adequate protection. In some rural areas, low water pressure, or even absence of a mains supply, can severely hamper fire fighting. The cause of this fire was use of the angle grinder in conditions where the sparks which are a normal part of its operation could come into contact with flammable materials such as saw-dust and volatile liquids. Had the '15-metre Rule' been in operation, this fire would not have occurred.

Welding poses the same type of fire hazard as cutting, in that particles of hot metal are spattered onto nearby surfaces. In general, there appears to be a better realisation of the fire danger associated with welding and of the need to clear the surrounding area of combustible materials. This may be because the equipment is unlikely to be used by persons other than properly trained operatives. Angle grinders on the other hand are freely available to handymen and are frequently used by persons with little training in their operation and none in fire safety.

Gas equipment should also be operated with extreme care, but some types are available to untrained operators. The oxyacetylene flame is clearly extremely hot and an obvious source of fire danger. However, oxygen and propane equipment is sometimes treated with less respect, as in the following instance, where a fire occurred in single-storey industrial premises being prepared for renovation. The electricity supply had been disconnected and subcontractors were using oxygen and propane gas cutting equipment to remove pipe-work from the building. The operation consisted of cutting off steel central-heating pipes where they emerged from the walls. Some of the pipes passed through wooden skirtings or other wood-product panels.

Two operatives were working overtime on a Saturday morning, removing piping from a corridor. They made several cuts, including one where the pipe emerged from a chipboard panel at the base of a cupboard fixed to the wall. One man cut the pipe with the gas torch, the other then cooled the cut-end by throwing water on it from a container, which they had with them for the purpose. They left the premises shortly afterwards. The charge hand stated that he checked the corridor before leaving the building and noticed no sign of smoke. A fire was discovered about 2 hours later, emerging through the flat roof in the area where the men had been working.

On investigation, it was clear that the outbreak had originated at the base of the cupboard where the pipe-cutting had been carried out. The cupboard base was completely burned away in the area where the cut end of the pipe emerged and the fire had spread up the side of the cupboard and to nearby timbers in a door and door-frame, before penetrating through the wooden roof-deck. Ignition of the combustible mineral felt that covered the roof carried the fire into other compartments before it was brought under control. When other pipe-cutting work which the same subcontractors had carried out on the site was examined, it was noted that considerable heat was transferred to nearby surfaces during the process. At the site of one cut, extensive smouldering had clearly occurred at the back of a wooden skirt-ing, dying away eventually for lack of fuel. The cooling measures taken following the pipe-cutting had clearly been inadequate and the men admitted having left the corridor site without maintaining the prescribed hour's fire-watch. Had they remained on site, it is unlikely that the fire would have developed undetected.

In recent years, a self-adhesive roofing material known as 'torch-on' felt has replaced the application of liquid tar as a method for waterproofing flat roofs. The felt is supplied in rolls measuring $1 \text{ m} \times 10 \text{ m}$, with a layer of bitumen on the underside. It is cut into suitable lengths and applied in overlapping strips to the roof-deck. Before each strip is applied, the bitumen backing is melted by application of naked flame from a propane gas torch. The strip is then pressed onto the roof-deck with a roller. In theory, the bitumen coating should remain soft for a sufficient time to allow the edges of the strip to be thoroughly fixed to the sub-strate. In practice, it is usually necessary to re-apply the flame along the edges of the roof to ensure that good adhesion is obtained, particularly at corners, under the overlapping edges of adjoining sloping roofs and on the masonry of adjoining walls. In such places, when the gas torch is applied to the underside of strips already partly stuck down to a deck, the flame is likely to encounter not only the bare wood of the deck but also timbers, roofing felt and accumulations of dust, leaves and other debris in crevices or under the edges of adjoining roof sections. This has emerged as a particularly hazardous operation, resulting in so many fires since the use of self-adhesive felt became common that the operation is now difficult to insure. Although some roofers will claim that the propane torch is never applied close to the roof surface, torch marks found on timbers during investigation has proved that this is not the case. The following incidents illustrate the range of fires which have resulted from the use of propane gas torches during roofing operations.

A typical case involved a large split-level bungalow, with single and two-storey sections. There were roofs at several different levels in the two-storey section, including an area of flat roof over the two dormer windows of the main bedroom. The main roof was tiled over tacked-down roofing felt in the usual way, but the flat roof consisted of a timber deck covered with mineral felt, waterproofed with brushed-on bitumen. On the day of the fire, a contractor was replacing the old felt on the flat roof with the self-adhesive type. The work had been started at the outer edge of the roof, and the felt laid in overlapping layers up the incline towards the main roof. The edge of the felt on the main roof was lifted to allow the self-adhesive felt to be inserted underneath it. While the gas torch was being used to melt the bitumen backing of the new felt, the edge of the felt under the tiles was accidentally ignited. The fire spread quickly up under the tiles, out of reach of the workman, who was then unable to control it. The fire spread to the entire central section of the roof and began to spread to other sections before it was brought under control. The newly felted timber deck survived largely intact, but the old felt under the tiled roof was entirely burned away and a number of the roof-trusses were destroyed.

A second case involved ignition of underlying timbers while self-adhesive felt was being applied to a flat-roofed extension. This incident occurred at a corner, where an extension to the rear of one of a terrace of houses adjoined the main building line at the level of the eaves, making right-angled junctions with the neighbouring houses on both sides. The original houses were brick-built and roofed with natural slate. The extension was roofed with a fibreboard deck finished with mineral felt . The felt was carried down over fascia boards at the edge of the flat roof and under the edge of the main roof to make a waterproof seal.

In order to stick down the edge of the roofing felt, the torch has to be applied close to the roof itself. The importance of getting a good watertight seal in this situation, and the difficulty of achieving it, can lead operators to overuse the gas torch in a place where it is virtually impossible to monitor the effect on the timbers underneath. It is at this stage of the operation that the greatest danger of fire exists, as timbers under the felt may become overheated and begin to smoulder. In the case illustrated, fire broke out in the timbers under the felt in the corner where the extension met the eaves of the original building. The fire appeared to have smouldered along the timber wall plate at the top of the original rear wall, spreading to the neighbouring house before developing to the point where the smoke began to issue from under the slates of the main roof. By this time the fire was spreading rapidly upwards and outwards in the roofs of both houses. In this old terrace, the party walls stopped short of the roof apex, leaving a tunnel through which the fire was able to spread to several more houses on both sides before it was brought under control.

The operator is often unaware of this type of fire until it has spread beyond easy control. In the early stages, smoke may be hidden within the roof cavity and it is not unusual for smouldering to continue for one or more hours before the fire develops sufficiently to become visible outside the roof. Some fires of this type could be avoided by appointing an operative whose sole function it is to keep a fire watch during the hot-work operation and for at least an hour continuously after it has been completed. However, there is often a reluctance to accept the susceptibility of timber to smouldering ignition and a tendency on the part of small-scale operators to dismiss the fire danger associated with this type of work.

As the following incident illustrates, failure to observe hot work protocols is not limited to one-man operations on single-habitation sites. It occurred on a roof-deck at the top of a nearly completed new city apartment building. The external walls were faced with brick over a central insulation layer of expanded polystyrene slabs. The concrete roof-deck was to be waterproofed by applying an outer covering of self-adhesive felt. A roofing contractor was employed to perform the operation, with instructions to continue the felt as a lining to the rainwater drainage channels running out through the base of the parapet. It did not occur to anyone to inspect the channels before the work commenced, so the roofers proceeded to use the gas torches, unaware that the parapet was a continuation of the main wall and that the channels were open to the polystyrene filling in the central cavity. The inevitable fire spread through the polystyrene, resulting in the virtual destruction of the outer wall of the building and an estimated cost to insurers of hundreds of thousands of pounds.

Mechanical failure

Mechanical failure of machinery can pose a fire hazard if moving parts become overheated through increased friction. Observation of recommended service intervals and periodic examination for wear and tear is essential for safe operation of all machinery, whether industrial or domestic. Machinery should be shut down immediately in the event of unusual sounds or odours during operation. It is also necessary to provide appropriate fire detection and containment systems, as the following incidents illustrate.

The first case involved a fire in a small packaging plant, which produced plastic bottles for the soft drinks industry. The premises consisted of a purpose-built factory building, divided into a production area and adjoining warehouse. Construction was mainly of sheet steel panels, supported on a framework of steel and reinforced concrete and finished with a PVC skin. The production area was divided from the warehouse, but the partition was not fire-proof. Polyester 'preforms', short bottle shapes, with threaded opening and thick walls, were fed into specialist moulding machines, where they were pre-heated and stretched and then inflated into the final bottle form with low-pressure compressed air. This was an automated process, run on a 3-cycle, continuous 24-h basis. Both raw materials and finished product were stacked as palletised loads in the warehouse and yard.

The compressor supplying the plastic moulders with compressed air was located at the inner end of the warehouse, close to the partition with the manufacturing area and within sight of a connecting door. The particular compressor model was chosen because it was designed for continuous 24-h operation and produced warm air as a by-product. It consisted of an air compressor and coolant pump, both belt-driven by a single electric motor. The coolant liquid was circulated through a radiator, where the heat was transferred to air blown through it by a fan, also operated by the electric motor. The compressor was capable of supplying up to seven moulding machines. The mechanism was protected by steel top and side panels and by a surrounding cage.

The compressor was about two years old at the time of the fire and was serviced at 6-monthly intervals. It was due for service early in January, and a representative of the service company had called about mid-December to book the service date. As the factory was in operation at the time, no inspection was carried out, but the machine was booked for the regular service and it was noted that the drive belt was due for replacement. During the night shift about one week later, one preform moulder was in operation, with the compressor running under automatic control to provide low pressure compressed air as required by the process. The machinery appeared to be running normally, with no sign of electrical or mechanical problems and the lights were functioning as normal, without flickering or dimming, when the duty manager noticed smoke coming from the door to the warehouse. On opening the door he found that black smoke was coming from the direction of the air compressor and he could see flames coming from the air vents, close to the belt-drive, fan and radiator. He sent a foreman to raise the alarm and attempted to control the outbreak with fire extinguishers, but could see that it was spreading rapidly to the warehouse contents. It became obvious that the plant should be evacuated and this was done, the manager switching off the electricity at the mains before he left. By the time the first fire tender arrived on the scene, the outbreak had spread into the factory floor, resulting eventually in the complete loss of the building and its contents.

Examination of the compressor control panel showed that all insulation and printed circuit boards were burned to ash but there was no evidence of particular 'hot spots'. No arcing or other evidence of electrical fault was detected and all the terminations were firmly in place. The evidence indicated that the fire had attacked the control panel after the electricity supply had already failed or been switched off.

All traces of the drive belt had disappeared. The aluminium casting on the front of the coolant pump (immediately to the left of the lower pulley) had melted and dropped down below the pulley. The fan blades and the radiator elements, both aluminium, had also melted and collapsed. Aluminium castings on top of the compressor had melted and dropped down on top of the lower pulley after it had come to a halt. Near the drive-belt area, the pipes of the cooling system were mostly disconnected, indicating that the heat in this region was sufficient to melt or burn the connecting seals. To the left of the control panel, most of the seals had survived, indicating cooler temperatures in this area. The motor windings and the inside of the front casting of the pump were intact, indicating that the fire had not been caused by seizure of the motor or pump.

Both the eye-witness evidence and the pattern of burning indicated that the fire had originated in the area of the belt-drive at the front of the compressor unit. Sudden shredding of the belt would cause snagging and friction, with the resultant heat leading rapidly to ignition of the belt fabric. With a mechanical rather than an electrical fault, the moving components can continue to function long enough for a self-sustaining fire to develop and begin to engage other combustible materials in the vicinity of the origin. Although it was known that the drive belt was due for renewal at the next service, the normal service intervals had been observed by both user and supplier and it appeared that nothing indicated that it was in imminent danger of failure.

However, general fire safety measures in the premises were less than adequate. In the absence of adequate fire-warning devices, the outbreak was well established before it was detected. Although the compressor was surrounded by a protective framework, this was intended only to avoid impact from pallet trucks and did not provide separation of the compressor from the warehouse contents. As a result, the fire spread rapidly to the plastics and cardboard stacked nearby. The structure of the building was completely inadequate to contain the fire in the compartment of origin, instead allowing it to break through within a short time to the production floor. Had a fire wall separated the storage and production areas, it would have provided the possibility of containment and total loss might have been avoided.

The second incident involved a 17-year-old domestic extractor hood unit. The householder had finished cooking a meal and had switched off the cooker, but left the extractor running to clear steam and cooking odours from the kitchen. She soon became aware of a spitting and bubbling sound, as if a pot were still boiling. At first she could not trace the sound, but then she looked in the direction of the extractor and saw a small flame running along the back of the filter. She did not remember seeing smoke at that stage. She immediately sent a member of the household to a neighbour's house to call the Fire Brigade, while she herself tried to place a fire blanket over the extractor unit to control the fire. She switched off the electricity and closed the kitchen window and door. As the fire had got worse, she evacuated her family from the house and held the fire blanket and another blanket which her neighbour had brought over the extractor unit until eventually she also was driven out by the thick black smoke which began to come from the burning unit. She left the house, closing the doors as she went. The fire brigade arrived soon afterwards and quickly extinguished the fire, which was still confined to the immediate vicinity of the extractor unit, although there was widespread smoke contamination.

The extractor unit consisted of a fan and motor mounted in a light steel case. A removable tray held a disposable polyester filter, mounted below the fan. The cooking vapours were sucked through the filter and blown out through a duct attached to a port at the back of the fan housing. The filter, which had been replaced one week before the fire, was intended to trap grease and dust, preventing them from sticking to the fan.

Examination of the extractor unit showed that the interior was very severely burned. The fan blades (probably plastic) had melted and burned and had dropped down onto the filter pad. The top of the extractor unit was distorted, showing that the fire had been inside the body of the appliance. In contrast, the only part of the filter that had ignited was immediately below the fan port, where burning material had fallen onto it from the fan. A ring of melted material surrounded the burned area. Beyond that, the rest of the filter was not merely intact but clean, indicating that the fire had not originated in the filter, or from the cooker or from food burning on the cooker. The inside of the flexible PVC duct which ran from the extractor to the external vent was smoke stained, indicating that the appliance had continued to work for a short period after the fire had started and the fan and duct had drawn off the smoke at the early stages of the fire. This suggested that the breakdown was mechanical rather than electrical.

When questioned, the householder recalled that she had switched the fan on and off several times, because the sound of the fan was 'different', and she was trying to 'clear' it. It appeared that the fan had begun to malfunction when it was switched on, but the householder persisted in using it, not realising the danger of doing so. The motor eventually overheated to the point where combustible materials such as insulation and other plastics ignited, but it continued to run until the fire cut the power supply. The appliance was unusually old, and appeared to have been well designed and robust. It had never required servicing or repair, but eventually had come to the end of its useful life. The incident illustrates the need for vigilance in relation to unusual sounds or odours during the operation of domestic electrical appliances. Although the householder succeeded in containing the fire while awaiting the fire brigade, she risked her life needlessly by remaining alone in the kitchen with the fire in progress. The correct action would have been to leave the house when she was satisfied that all members of her family had been accounted for.

In the third example in this section, a design deficiency resulted in an outbreak of fire during an automated industrial process. The incident occurred in a small factory unit, which manufactured customised moulds for industrial processes. The fire was confined to a single machine, a spark erosion die-maker, but caused extensive smoke and heat damage, resulting in loss of production for several weeks.

The spark erosion apparatus operated by using an electrode to deliver a series of electrical discharges (sparks) to a target consisting of a steel blank. The sparks eroded the surface of the steel to form a cavity of the required pre-programmed dimensions. It was designed for automatic operation, switching off when the pre-set task had been completed. The spark discharge necessarily created a very high temperature, which was controlled by carrying out the process in a bath of suitable liquid, in this case a dielectric (electrically non-conducting) oil of high boiling point. The spark erosion process threw off metal particles, which had to be removed from the work-face and later filtered out of the liquid. This was accomplished by circulating the liquid so that a jet constantly washed over the target and electrode, and recycling it through a cooling and filtering device. The installation therefore consisted of three components:

1. the machining unit consisting of a moulding head, incorporating the electrode and a work tank for the coolant liquid, mounted on a calibrated, movable, platform;
2. the dielectric liquid unit incorporating cooling, filtering and pumping mechanisms to control the liquid temperature, composition and flow rate through the tank;
3. the generator unit, incorporating control and monitoring systems.

The dielectric coolant used was a petroleum hydrocarbon with the following characteristics:

Density	0.8 (i.e. lighter than water)
Distillation range	205–260°C
Flash point	80°C

This is a widely used dielectric coolant, which is safe under normal conditions. It does not carry the risks to health and environment of many of the non-flammable alternatives, such as the PCBs. Liquids of this type are used as industrial coolants because of:

1. their high distillation temperature – there will be little or no evaporation from the open tank during use;
2. their high flash point – when recycled through a cooling mechanism the temperature will remain well below the critical level; however, safety depends on maintaining the liquid at the correct level in the tank;
3. their insulating properties – they do not conduct electricity.

At the commencement of a machining operation, the levels of the work tank and the moulding head were adjusted to suitable starting positions, the former manually and the latter automatically. The level of the dielectric liquid was set to provide a depth of 25–50 mm of fluid over the working surface. The liquid level was set by manually adjusting the height of a sliding panel, which controlled the return overflow. This was done by drawing up a rod, which pulled up the panel until the top was at the desired working level for the liquid. The rod was then secured in position by screwing in a second rod to hold it in place. The liquid was then pumped into the tank from the dielectric unit until it reached the overflow level at the top of the sliding panel.

A float switch was attached to the sliding panel to monitor the level of the liquid in relation to the panel top. During operation of the spark erosion process, the liquid was pumped in through a hose attached to an inlet valve at the top of the tank and recycled through the cooling and filtering processes in the dielectric unit. If the liquid level dropped below the sliding panel-top the float switch should operate to shut down the machine.

On the evening of the fire, the spark erosion machine was set for automatic working just before the staff left at 17.00 h. The machining operation was due to finish at 23.30 h, but at about 22.45 h security personnel discovered a fire in the workshop. The premises were found to be secure and unoccupied when the emergency services reached the scene.

Examination of the fire scene indicated that the outbreak had originated in the spark erosion machine and was confined to it and its immediate vicinity. It was clear that the surface of the liquid in the tank had ignited. The machining unit was severely damaged, but it appeared that the fire had not occurred as a result of an electrical fault in the wiring in the head. There was no evidence of electrical arcing and in any event the small quantities of hot debris which such a fault would produce would be insufficient to ignite the liquid in the tank, provided it was at the correct working temperature. Failure of the cooling mechanism would have caused the machine to shut down automatically, so it was considered unlikely that the fire was caused by such a failure. The only source of ignition hot enough to ignite the cooled liquid was the electrical discharge between the electrode and the target work piece. This could only occur if the surface of the liquid fell to the level of the spark discharge without triggering the cut-out mechanism.

The float switch cut-off should in theory have prevented operation of the machine unless the liquid was at the pre-set level. However, because the float switch was attached to the overflow control panel, it actually monitored the liquid level only in relation to the position of the panel-top and not in relation to the working surface. It was discovered that the device holding the sliding panel in position at the pre-set level was a simple screw, pressing against the positioning rod. There was no mechanism to monitor the position of the panel in relation to the working surface. Loosening of the screw, due to wear or insufficient tightening coupled with vibration during operation of the machine, could conceivably have allowed the rod and panel to slip down, in turn allowing the liquid to drain out through the overflow. The float switch, attached to the panel, would sink at the same rate, so that the falling liquid level was not detected and the safety cut-out would not operate. This clearly could bring the liquid surface to the level of the spark discharge and lead to ignition of the coolant liquid as described here. It was concluded that the accident occurred because the method of monitoring the coolant level in the tank and the method of maintaining the level were not independent and were both inadequate.

Overheating of electrical installations

Heat is generated when an electric current passes through a conductor and provision must be made for its safe dissipation. Appliances are provided with heat sinks and ventilation grilles, and instructions-for-use warn that these should not be covered or obstructed. However, it is often not realised that plugs, sockets, fuses and lamp holders, while continuing to function normally, can become hot enough in certain circumstances to constitute a fire hazard. This is particularly so where these items are in contact with wood or wood products, as smouldering can be initiated in wood at temperatures which are far lower than those required to ignite the plastics used to manufacture electrical components. Plugs and sockets used to connect built-in dishwashers and clothes dryers and hidden under kitchen worktops or behind the backs of the appliances are particularly at risk, due to the relatively high current (close to 12 A in some cases), drawn by these appliances when the motor, pump and heating element are operating together, as in a clothes drier and in the rinsing cycle of a dishwasher. This is because the flat pins of the ordinary 13 A fused plug with which many such appliances are fitted, do not make sufficiently good contact within the socket terminals, leading to the development of 'hot spots'. Eventually the plug may disintegrate, or the internal insulation in the socket may break down, posing a serious fire hazard. In addition, the hot socket may ignite the surface to which it is attached, if this is combustible. The socket may become hot enough to cause wood to char long before the plastics become hot enough to burn. This sequence of events can occur without rupturing the plug fuse or tripping the MCB or RCD, because it happens at normal current loads and without causing a short circuit.

An unusual smell provoked one householder to investigate the electrical connection to a dishwasher recently installed in her kitchen. It was found that the 13 A fused three-pin plug had become discoloured over the position of the plug fuse and that the surface of the socket was discoloured around the live terminal. A new plug was fitted and a different socket was used, but the smell occurred again within about three months. The new plug was found to have become discoloured in the same way and overheating had extended from the 'live' pin of the plug, along the plug fuse and into the 'live' conductor in the flex. It had also affected the 'live' terminal of the socket, and the associated wiring. The parts of the plug close to the fuse and 'live' pin had become charred, as also had parts of the socket. Left undetected, such a situation could clearly lead to fire. In this case, the plug and socket connection was replaced with a permanent connection through a secured junction box to an isolator switch above the work-top. The appliance functioned safely for the rest of its projected life span.

The smell of charring urea formaldehyde plastic resembles that of rotting fish. This is usually the first indication of an overheating plug or socket, but may not be noticed by a householder if the machine is installed in an area such as a utility room, as in the following case, where a fire occurred in the utility room of a modern house. The room was fitted out as a laundry, with a sink, washing machine and tumble drier. The tumble drier was some years old, and the washing machine about two months in use. The two appliances were installed under a work-top, where both were plugged in to a double socket fitted below the back of the worktop. In order to disconnect the appliances from the power supply, or to inspect the plugs and socket, it would have been necessary to pull the appliances out from under the worktop.

On the morning of the fire, the householder had switched on both the washing machine and the tumble drier and gone out to play tennis, leaving the appliances to finish their cycles and switch off automatically during her absence. She returned about 2 hours later and was in the kitchen when she heard a hissing and crackling sound, which she described as like timber burning. Then she heard a bang from the utility room and went to investigate. She found flames running up along the wall at the back of the work-top, above the washing machine and drier. She raised the alarm, but the fire had spread to the roof over the utility room before it was brought under control. An examination of the remains of the work-top showed that it was burned completely away in the area over the backs of the washer and drier. The two appliances were still relatively intact and the fire clearly had not originated in either. The double socket had been attached to a panel, which formed the back of the built-in unit, rather than directly to the wall. The remains of the unit were recovered, still attached to the wiring, with all terminations firm. Except for the metal parts, the double socket and plugs were totally destroyed.

Modern practice is to install sockets for built-in domestic machines in a cupboard beside the appliance, where they are accessible, but many of the older installations are still in use. It would be safer to connect built-in washing machines, dish washers and tumble dryers, fitted under worktops, into isolating switches mounted above the worktop and incorporating warning lights. The connection can be made via a junction box fixed below the level of the worktop so that the appliance can be easily disconnected for maintenance or replacement. This would avoid the use of plugs and sockets altogether, and lessen the risk of overheating and fire involving such appliances.

Also potentially dangerous are recessed lights ('down-lighters'), unless care has been taken to locate them where there is plenty of clearance between the back of the lamp and the nearest timbers. Handymen or home decorators in particular may install low-voltage systems without understanding the need for adequate space and ventilation, but poor quality installations can be found anywhere.

A new conference and banqueting wing had been added to an established hotel. It was modern in design, with a suspended mineral slab ceiling in which a low-voltage recessed lighting ('down-lighter') system had been installed during construction. A transformer with plug-in connections was provided for each lamp, and wiring and transformers were safely accommodated in the ample space above the suspended ceiling. The new wing was accessible from the main entrance through the original single-storey foyer. The foyer had a flat roof, whose construction included a chipboard deck laid on an upper set of 120×50 mm timber joists, which in turn rested on a lower set of 180×50 mm timber joists (with some RSJs). Glass-fibre insulation felt was in place between the roof and the lower timbers, to which a double plaster-slab ceiling was pinned. A maximum clearance of 180 mm was therefore possible between the ceiling and the glass-fibre insulation layer.

When the new wing was completed, it was apparently decided to install matching recessed lighting in the foyer ceiling. Holes were cut at regular intervals in the existing ceiling slabs, continuing the arrangement in the linking corridor of the new wing. The new wiring and transformers were inserted into the space between the glass-fibre matting and the ceiling and the down-lighters were pushed up into the prepared holes, where they were secured by integral clips. The lights in the foyer were in almost constant use, being switched on for practically the entire 24 h of every day. They were controlled in groups by switches near the reception desk.

One day some months later, a duty manager noticed that some of the recessed lights near the entrance to the new wing were not lighting. She checked the switches and found that they were in the 'on' position. She next became aware of a noise, which she thought sounded like dripping water. She sent another staff member to investigate the noise and check for a leak. The second woman could hear the sound near the entrance to the new wing, but did not find any water. She noticed instead that the light nearest the entrance to the banqueting wing was flickering red. The women then noticed a peculiar smell in the area, and finally smoke began to drift down from the ceiling. Eventually realising that she was dealing with a fire, the duty manager raised the alarm. By the time the Fire Brigade arrived, the fire had spread across the foyer roof to the roof of the new extension.

The duty manager was precise in her placing of the first indication of the fire. The burning patterns in the area that she indicated supported her belief that the fire started in the vicinity of the light where her colleague had seen the red flickering. This may actually have been flames, seen through the lamp.

When the area was examined, it was noted that two rows of down-lighters near the entrance to the new wing coincided with the edges of two of the timber ceiling joists. Deep sockets had been chiselled into the joists to accommodate the tops of the lamp holders. The surface of the wood inside one of the chiselled sockets which had not been reached by the fire was found to have been darkened by the heat of the lamp and a projecting splinter which had been closest to the lamp was charred black.

Unlike the installation in the new wing, which had been planned for at the design stage and was standardised and well ventilated, the installation in the ceiling of the foyer was added as an afterthought and was not standardised. Insertion of the lamps into hollows gouged in the timbers above the ceiling meant that the fittings not merely lacked proper ventilation, but were probably in contact with the roof timbers. This situation led directly to a fire that resulted in destruction of property and interruption of business and might have had considerably more serious consequences if it had occurred at night.

Spontaneous combustion

Exothermic chemical reactions can occur at ambient or even low temperatures, resulting in physical and chemical changes in the materials involved. Oxidation of plant and animal materials is the most widespread, but not the only, reaction of this type.

In some instances, oxidation takes place as a more or less direct chemical change in the substrate, such as oxidation of metals or of unsaturated chemical bonds in fatty acids. Often the natural oxidation reaction is slow, with imperceptible rise in temperature, but with noticeable physical and chemical effects. Metals develop a coating of oxide, cellulosic materials darken in colour and may become brittle and edible fats develop bitter or rancid flavours. In many natural processes, the oxidised outer layers become a desirable protective coating for the rest of the mass. The condition may then be described as 'patinated' in the case of metals or 'weathered' in the case of timber.

As already described, long-term contact with hot water or steam pipes operating at well below the ignition temperature of wood and wood products may induce the production of charcoal [1]. A fire apparently involving this phenomenon occurred in a nineteenth-century building housing a community of religious sisters. The fire was discovered at 1 a.m. when smoke was noticed in a first-floor corridor and was traced to an unoccupied guest bedroom. The room had last been occupied by a visiting sister, who stayed for two days and left at 12 noon on the day before the fire. The room was then cleaned and the door was locked. Neither the visiting sister nor the sister who cleaned the room were smokers. The building was heated by an oil-fired hot-water system, installed at least thirty years previously and consisting of large cast iron radiators fed by steel pipes of approximately 20 mm diameter. The boiler was operated at maximum temperature, close to boiling point, for some hours each morning and from 5 p.m. to 9 p.m. each evening. The floor of the room in which the fire occurred consisted of plane-edged planks of American pine ca. 20 mm thick, supported by 230 mm × 50 mm joists and overlaid with hardboard and a heavy felt-backed PVC floor covering. The radiator stood against the external wall, supported on two 100 mm steel brackets, bolted to the floorboards. The hardboard and PVC were cut to fit around the brackets. The room was simply furnished with bed, locker, chest of drawers, wardrobe and wash basin. There were no electrical fittings or appliances other than a central ceiling light operated by a switch on the wall inside the door and a single socket on the same wall, which separated the room from the corridor. The window was intact and the door of the room was still locked when the fire was discovered. As far as could be ascertained, no one had entered the room for at least 11 h before the fire was discovered.

The fire appeared to have started in the floor around one of the radiator brackets. Two joists, one on either side of the bracket were burned away for a distance of 2 m from the external wall, while the floorboards above them were burned away for a distance of less than 1 m from the external wall and were charred on the underside for about a further metre inwards from the wall. The hardboard and PVC were charred through only in the area beside the outer wall, where the skirting was also burned away (Figures 2.1 and 2.2).

The floor covering and the charred floorboards remained in place until removed by firemen. The contents of the room were intact, except for being completely blackened by smoke, an indication of the long duration of the outbreak before it was discovered. At first sight, the fire damage appeared typical of that caused by an under-floor electrical fault, but the only wiring, which was still completely intact, led to the ceiling light of the room below and did

Figure 2.1   Convent fire viewed from beneath the fire room

Figure 2.2   Convent fire view showing the area near the radiator

not come within a metre of any part of the floor affected by the fire. The ceiling of this room was lath and plaster and the 15 mm-thick plaster was still intact when the fire was discovered. It is difficult to come to any conclusion other than that the fire was the result of pyrolysis over a long period, of the floor timber to which the radiator support had been bolted for over 30 years.

In many cases, oxidation occurs first as part of the respiration process carried on by living organisms, following which heat is transferred to other materials in the surroundings. The overall respiration reaction in aerobic conditions is the stepwise oxidation of D-glucose according to the following general equation:

$$C_6H_{12}O_6 + 6O_2 \rightarrow 6CO_2 + 6H_2O + \text{energy}$$

The energy released drives the life processes of the organism, both by fuelling further chemical reactions and by maintaining the temperature at the optimum level for chemical and physical activity. This is usually about 37–38°C. Under normal environmental conditions, living organisms control cell temperature by evaporation of water from internal and external surfaces, a process requiring adequate ventilation. In the absence of ventilation, the cell temperature rises. Above about 40°C, the cells of most green plants and the majority of animals quickly die, but thermophilic micro-organisms flourish and continue to proliferate up to temperatures of at least 60°C. With few exceptions, the proteins of cytoplasmic systems begin to coagulate between 60°C and 70°C, resulting in breakdown of the cell and death of the organism.

In order to ensure the preservation of harvested material, methods have been developed to slow biological activity by reducing either the moisture content or the oxygen content before or during storage. Air-drying is used for the preservation of bulk plant materials such as fodder, grain and other seeds and fruits. The amount of water that can be removed from organic material by drying varies from about 50–90% of live weight. Fresh, leafy grass can have a moisture content of up to 70–80% by weight, while animal tissues may have even higher moisture contents. Producers aim to harvest grain at a moisture content of not more than 17% by weight. This is further reduced to a safe level of 10–15% by drying in silos through which air at suitable temperature and humidity is drawn. Inadequate drying can produce conditions favourable to fungal, bacterial and insect infestation, leading to spontaneous heating during later bulk storage.

The spontaneous combustion of haystacks has been a subject of interest and research for many years. Cut grass is spread to dry before stacking or baling in the field, where it is then allowed to stand for a further 'curing' period to ensure that all biological activity in the grass cells has ceased before final stacking in open-sided barns as hay. Even in the absence of external moisture, the water vapour produced by respiration in inadequately cured hay can favour the proliferation of fungi and bacteria. Heating can also follow re-wetting of stacked hay, for instance following leaks or flooding. Production of steam is usually the first sign of spontaneous heating, which can occur at any time from about 10–90 days after stacking.

Rothbaum [9] concluded that in the spontaneous heating of hay, microbial action, in the presence of adequate oxygen and moisture, is responsible for heating up to about 70°C. The temperature may then be further raised by exothermic chemical reactions which are still dependent on oxygen and moisture balance, but do not require the presence of breakdown products derived from previous microbial action. Beyond 100°C, further chemical oxidation of dry hay can heat it to ignition temperature. Currie and Festenstein [10] demonstrated that the physical processes affecting self-heating of hay between 70°C and 100°C are heat balance, aeration and moisture movement. The centre of a haystack tends to become anaerobic, but heating to ignition can occur in channels leading towards the outside of the stack, where the hay has dried and air can penetrate. Only when the temperature exceeds 100°C does heating tend to proceed at an increased rate.

Many instances of spontaneous combustion of baled hay were seen in Ireland during the early 1980s, before the preservation of fodder in the form of silage became more popular. The scenario was almost always the same: during a spell of dry weather, hay was cut and baled after a few days, as soon as it appeared dry enough. The bales were then stood on end to dry further before drawing into storage. A threatened break in the weather induced an earlier completion of the task, and the inadequately dried hay began to heat in storage. Counts of thermophilic bacteria per gm dry weight of the damaged hay were typically found to be in excess of 1×10^6. If it was noticed in time, remedial action could be taken, but in some cases a rapidly developing fire took owners by surprise, resulting in loss of buildings and livestock and machinery as well as the crop. The shortest time recorded in my experience from harvesting to fire was 12 days. Following a long spell of dry weather, the hay had been cut, baled, and drawn in within 48 h. In this case, the grass was externally very dry because of the prolonged drought, but clearly the inner cells were still alive when it was baled and stacked. Ten days later it burst into flames in the middle of the night, resulting in the loss of four horses in an adjoining stable as well as the barn.

Spontaneous combustion of hay mainly involves the oxidative degradation of starch and cellulose and their break-down products, as non-reproductive plant tissues contain very little fat. However, seeds of all sorts, which include nuts, grains and pulses, store energy in the form of oils as well as starch and sugars. The oils consist mainly of esters known as glycerides, formed by the linkage of three molecules of fatty acids to a molecule of glycerol. Mixed triglycerides predominate in nature, with two or three different fatty acid radicals linked to a single glycerol molecule. The natural fats and oils contain a relatively limited number of fatty acids, several of which are unsaturated, that is, containing carbon–carbon double bonds. These substances can undergo direct oxidation on exposure to air at room temperature. Rothbaum [9] notes that in laboratory adiabatic experiments into spontaneous heating of plant materials, investigators recorded a temperature of 101°C with moist soya beans and 94°C with moist wheat, both well above the level attributable to microbiological activity.

The degree of unsaturation is indicated by the iodine number or iodine value, a number expressing the percentage of iodine absorbed by the substance. The most highly unsaturated fatty acid is arachidonic acid, with an iodine value of 333.50. It is an essential fatty acid associated with vitamin F and containing four carbon–carbon double bonds. It occurs in liver, brain, glandular organs and adipose deposits and is a constituent of animal phosphatides. Linolenic acid, which contains three carbon–carbon double bonds, is found in plants. It gets its name from linseed oil, of which it is an important constituent. Linseed oil has an iodine value of 175–205 and is one of a number of highly unsaturated plant oils (also including tung oil and perilla oil) described as drying oils, which tend to dry and harden on oxidation, and are widely used as constituents of putty, polishes, lacquers and paints.

The most abundant polyunsaturated fatty acid is linoleic acid, an essential fatty acid that is a major constituent of many plant oils and is also present in animal fats. The molecule contains two carbon–carbon double bonds and is a characteristic constituent of unsaturated (semi-drying) oils, including soya bean oil, maize (corn) oil, sunflower oil, rapeseed oil and cottonseed oil. Fish oils including sardine, herring, cod and shark liver oils also fall into the category of semi-drying oils. Oleic acid, the molecule of which contains one carbon–carbon double bond, is a major constituent of tallow (animal fat) and also occurs in plant oils. Fats and oils containing linoleic and oleic acids are major food constituents and are also used as cooking fats. Materials containing or contaminated by the semi-drying oils are prone to spontaneous heating, especially if packed or stored in conditions above ambient temperature. Almond, olive and castor oils are non-drying oils, containing fewer unsaturated linkages, and are less likely to self-heat.

The unsaturated fatty acids are constituents of various products assembled in bulk. These include plant residues of other manufacturing processes, destined for incorporation into animal feed (e.g. carob shells, sugar beet pulp) and animal products and by-products such as milk powder, fishmeal, hides and raw wool. All are prone to spontaneous heating unless correct moisture content and temperature are maintained in storage. Meat and bone meal, once a common animal food, is less likely to be encountered in general storage since the adoption of incineration of animal waste as a disease prevention measure, but is encountered at waste management facilities, particularly rendering plants. Oils and fats such as tallow (animal fat) and waste cooking oil, which are solid at ambient temperature, are stored in tanks heated by steam pipes. Spillage onto pipe-lagging can result in spontaneous heating. The following examples illustrate the range of materials and situations in which spontaneous combustion involving fatty acids may be a cause of fire.

A fire occurred in the hallway of a three-storey house converted to small apartments. Inside the hall door a set of consumer units had been installed for the electricity supplies to the various tenants. Each consisted of meter and main fuse, with a set of trip switches and other circuit breakers. Six meters, seven fuses, the trip switches and some other units were installed in a timber cabinet, fitted with two timber doors. Six of the fuses were installed in a plastic box at the centre top of the cabinet. The meters were arranged below in a row of four and a row of two. The trip switches were placed near the top left of the cabinet, with timers for the heating system above them. There was space at the bottom of the cabinet, in which the landlord was in the habit of storing various materials. At the time of the fire, these included papers, cotton cloths and a quantity of fresh putty wrapped in plastic film. The cloths and putty had been placed in the bottom right corner of the cabinet, following replacement of a glass panel in the hall door eleven days before the fire. This was discovered one afternoon when smoke was noticed coming from the meter cabinet and the Fire Brigade were called to the scene. The burning extended in a fan shape from the bottom right corner of the cabinet to the top centre, burning into the right-hand side of each meter. The plastic cover over the fuses melted, leaving the fuses hanging down over the tops of other units. The timber was charred fairly evenly, with no evidence of intense heating associated with the position of the fuses. There was no indication that any of the fuses or other electrical components had caused the fire, which had not developed beyond the smouldering stage. The effects of a smouldering fire could clearly be traced upwards from the bottom right-hand corner of the meter cabinet, where the putty had been left. Investigation of the charred materials in the bottom of the cabinet revealed the presence of the still-fresh putty, wrapped in the remains of the sheet of plastic film, and surrounded by the charred debris of pieces of cotton towel. When examined following the fire, the putty was still leaking linseed oil. In the corner of the cabinet there would have been a sufficient air supply for oxidation to proceed, without adequate air movement to dissipate the heat produced – classical conditions for spontaneous combustion.

The following incident involved a fire on pipe-lagging. The premises consisted of a modern animal feed mill where various ingredients were combined to produce a range of animal foodstuffs. The ingredients included reclaimed cooking fats, consisting of a mixture of purified and clarified animal and vegetable fats and oils semi-solid at ambient temperatures but liquid at the handling temperature of 50°C. The fats were delivered by road tanker and pumped into large storage tanks standing in the open. The pipes used for filling and emptying the fat tanks were electrically heated, but the tanks were heated by steam pipes that surrounded the vessel and maintained the contents at 50°C. Steam was generated in a boiler house nearby and pumped to the tanks. A load was delivered at about 20.00 h (8 p.m.) on the evening before the fire. The evening shift finished and most workers left the premises at 22.00 h (10 p.m.). At that time, there was no evidence of anything amiss at the tank site. The fire was discovered at 04.00 a.m. by the night watchman on his rounds. He saw smoke coming from underneath the tanks, from the region where the insulated steam pipes ran under the vessels and realised that a fire was in progress. He raised the alarm and tried to control the fire with fire extinguishers. However, the fire appeared to have developed fairly rapidly, causing the filled tank to rupture and allowing burning fat to spread over the site. All four tanks on the site suffered major damage and the two fat tanks suffered partial collapse. The flames were also drawn up along the outside of the main building, entering the structure through a full-length panel of perspex windows.

There were no electrical installations in the area where the fire originated. The watchman was a non-smoker and there was no one else in the area for a number of hours before the fire. The pipe-lagging consisted of mineral fibre covered by plastic wrapping to protect it from spills. It is of the utmost importance that lagging is kept protected from spills by attention to maintenance of the covering material. However, it appeared the wrapping had been allowed to become tattered, so that protection was no longer complete. The reclaimed fats containing; saturated and unsaturated vegetable and animal fats and oils, subject to spontaneous combustion, had therefore been able to contaminate the lagging fibre. Although the contents of the storage tanks were kept at an average temperature of 50°C, the temperature of the steam pipes would have needed to be much higher in order to maintain this average. If the lagging on steam pipes becomes contaminated as a result of oil spills, there is a serious risk of spontaneous combustion, which appeared to have been the cause of this fire.

Another lagging fire occurred in a waste recycling facility attached to a paint factory. Recycling involved distilling white spirit from paint residue by heating the waste to a suitable temperature and collecting the condensate. Waste treatment was carried out using a patent still. The waste to be treated was placed in an insulated tank, mounted in the open at the rear of the treatment unit. It was heated by piping hot cooking oil, electrically heated in a separate insulated tank, through a coil in the waste tank.

The oil heating process was designed to operate at atmospheric pressure (although the tank had been tested to higher pressures). In order to maintain atmospheric pressure, a vent pipe was run from the top of the tank to the outside of the building.

The oil was Castrol 'Perfecto HT5'. This was described as cooking oil with the following parameters among its specifications:

Flash point cc	207°C
Flash point oc	227°C
Fire point	249°C
Ignition Temp	420°C
Boiling Point	360°C

The specification included an instruction that the oil was to be heated in the absence of air, as it was rapidly oxidisable at temperatures above 100°C.

According to the operators, the process was normally run at between 175°C and 220°C, and usually at about 185°C, well within the tank design temperature of 288°C. The requirement of the oil specification, that heating be carried out in the absence of air, would have been met by the design of the oil-heating tank, which was closed except for the vent pipe. The space above the liquid would rapidly fill with oil vapour as the temperature rose, expelling any air through the vent. A positive pressure would be maintained by vaporisation until the oil had cooled down again, thus excluding air from the tank while the oil was hot. The practice was to run the distillation plant for about five days at a time, then shut down for maintenance, which entailed clearing the outside of the oil-filled element of paint sludge. The level of oil in the heating tank was tested by dipping before it was used again. The tank was kept filled to within 150–200 mm of the top, maintaining a depth of at least 200 mm over the element.

The purification plant had been run intermittently over a number of years without incident, but immediately before the fire it had been disused for a period before being brought back into service. The oil level was checked by dipping before being brought into use about five days before the fire. It was normal for bubbling sounds to be heard from the tank during heating, once the temperature went above about 170°C (similar sounds can be heard from deep-fat fryers during heating). This occurs because the element at this stage is always hotter than the surrounding oil, and bubbles of vapour form on its surface. The process normally produced smells of solvent and hot oil while distillation was in operation. Nothing unusual was noticed during the days while the waste recovery cycle was running. One of the operatives noticed that the oil tank was slow to cool on the afternoon before the fire, and that bubbling sounds continued for longer than usual. This could indicate that a hot spot had developed in the insulation, due to spontaneous heating having begun in a small area affected by drips of oil following dipping of the tank. However, to someone who was not aware of the susceptibility of the cooking oil to spontaneous combustion under the circumstances described, there would be no obvious reason to suspect that there was any danger of fire, particularly as the general temperature in the tank continued to fall.

The fire was discovered on the outside of the tank at 02.00 a.m., some seven hours after the heating element had been shut down. Flames were also coming from the vent pipe, as the contents of the tank boiled off and ignited in the open air. The coolant in the drum just in front of the oil-heating tank did not appear to have become involved in the fire although electrical and pipe-work connections between drum and pump were disrupted. The drum was still full when inspected after the outbreak. The fire did not appear to have been electrical in origin. The electrical installation appeared rather to have been damaged by fire that originated externally. There was no evidence of intrusion.

The fire could not have originated inside the oil-heating tank. The flammable limits for most cooking oils lie in the region between 1% and 8% by volume of vapour in an air/vapour mixture. Outside this range, combustion is impossible. The spontaneous ignition temperature of the oil was well above the boiling point, so that all air would have been expelled from the tank before auto-ignition temperature was reached. The requirement that the oil should be heated in the absence of oxygen was adequately met by the design of the oil-heating tank. Once the oil in the tank was affected by an external fire however, it boiled off at high temperature, well above the point at which flaming ignition could be sustained (the fire point). The vapour emerging from the vent pipe mixed with the external air and could therefore be ignited by the sparks from the fire in the building. As the Fire Brigade poured water on the tank, the effect of cooling the vapour inside would be to decrease the pressure, leading to even more furious bubbling of the remaining liquid oil (this can be demonstrated in the laboratory), so that eventually all the oil vaporised and was burned off at the vent.

Research, carried out at Borehamwood Fire Research Station [11] has confirmed that spontaneous combustion is likely to occur in situations where mineral pipe-lagging or other inorganic insulation is exposed to heat after being contaminated with oil. It was found that the amount of oil was not critical and that spontaneous heating developed after an initial induction period, ranging from hours to days, during which the lagging was maintained at temperatures above a minimum of about 100°C. The amount of lagging affected by spontaneous heating could be quite small, with no obvious external effects until glowing char developed, and once initiated, spontaneous heating could continue after the initial heat source was removed. Although insulation was necessary to conserve heat in the initial stages, once heating was strongly established, it was favoured by the presence of air currents in the environment. In the case of the fire described here, the original cover over the tank insulation had been disturbed and only partially replaced. It is likely that small spillages occurred during the dipping operations, leading to contamination of the tank lagging.

In laundries, the temperature at which cotton articles emerge from dryers or rotary irons is often close to that of boiling water. As the hay research has shown, if cellulosic materials are packed at temperatures above 70°C in the presence of other factors which predispose to spontaneous heating, then a real danger of spontaneous ignition exists. There are many anecdotal reports of fires in laundries due to spontaneous ignition. The following are two incidents in which warning signs of possible spontaneous ignition were ignored. Both refer to cotton fabric contaminated with plant oils.

A laundrette offered a service to local businesses, whereby towels, cleaning cloths, etc. could be left in for washing and drying by the attendants. An employee had laundered and packed some cloths for a nearby bakery during the morning of the fire, finishing just before going to lunch at 1.30 p.m. She locked the shop, but left the side door to the self-service area open to facilitate lunch-time customers. The lights and water-pump were left on. This was the normal procedure. At about 2.15 p.m., a fire was noticed in the shop. The proprietor was telephoned and was the first to arrive on the scene. He came in through the side door and noted that the fire was confined to the shop area. There was no smoke in the launderette at that stage. The door in the glass partition separating the launderette from the shop was still locked as the attendant had left it, as was the street door of the shop. He opened the door in the glass partition with his keys to enter the front of the premises, which was full of smoke. There was a glowing fire in the bags of laundry left ready for collection. He attempted to fight the fire with the extinguisher which he had brought with him, but he failed to put the fire out and it appeared to him that he only succeeded in scattering the burning material and spreading the fire, which blazed up and involved more of the laundry bags. He ran to the back of the launderette to get another extinguisher, but was unable to re-enter the shop because of the smoke and heat. The fire was eventually extinguished by the Fire Brigade.

The fire did not appear to have been electrical in origin. There were no electrical appliances or connections in the area where the proprietor first saw the fire. As he came through the launderette, he heard the hum of the water-pump and as far as he remembered the lights were still on, indicating that the fire had not at that stage affected the electrical wiring. The attendant admitted when questioned that she had noticed 'fuming' coming from one of the bags while she was packing the laundry. She had ignored it and gone to lunch, assuming that it would stop.

The second premises consisted of a purpose-built factory unit in which wiping cloths of various materials were manufactured and/or stored. One product was a cotton knit fabric, suitable for cleaning industrial machinery. It was machine knitted on the premises from unwashed cotton fibre, produced in long lengths that were then cut into suitable shorter pieces before being washed and dried in industrial (laundry-type) machines. The purpose of the washing was to remove contaminants and loose fluff. The cloths were then packed, still hot, into cardboard cartons for dispatch. Each carton contained 10 kg and measured about 500 mm³.

On the day of the fire, some 60 kg (6 cartons) of knitted cotton cloths were washed, dried and packed. The cartons were then taken to a loft to await dispatch. The worker who usually carried out this operation was about to leave the company and was training his successor. He went home early in the afternoon, leaving the inexperienced worker to complete the task. According to the experienced man, smouldering had occasionally been encountered in the past at the drying stage of the process. The cloths were packed into the cartons straight from the drier and were quite hot. The factory was closed at the normal time that evening. All machinery and lights were switched off before the staff left.

The fire was discovered at 10 p.m. when smoke was seen coming from the building. The fire started in the cartons of washed cloths, which were stored in the corner of the loft. The cartons, which were standing on a pallet, and their contents were almost completely burned out and the fire spread to other goods. The Fire Brigade found the premises secure when they arrived on the scene. Smoking was forbidden on the factory floor and the worker concerned would have been clearly visible both in the washing and packing area and in the loft. No lights had been left on and there were in any case no electrical appliances or installations in the vicinity of the cartons.

The cloth was knitted from waste cotton fibre, examination of which suggested that it contained natural impurities, which would include plant oils. The washing process was mainly intended to remove excess lint or fluff and the temperature was probably not high enough to remove all the oils. Contamination by plant oils, which normally contain a proportion of unsaturated oils, would enhance the likelihood of spontaneous combustion when the cloths were packed hot from the dryer. The revelation that 'smouldering' or fuming had occurred previously during the drying procedure is an indication that spontaneous combustion was a real hazard in this process. The inexperienced operator had probably never met this situation and would not have recognised it as a danger.

While spontaneous heating of manufactured or partially manufactured meat and bone meal is relatively common, particularly when it is allowed to accumulate in large quantities, heating of the raw materials, that is, raw macerated animal tissue, is relatively unusual. However, this is what appeared to have occurred in the next incident.

The premises consisted of a large modern facility for the rendering of animal remains. The plant had been in operation for about 11 years and buildings and machinery were up to date and appeared to be clean and well maintained.

The raw material taken in included whole carcasses as well as remains from abattoirs, and butchers. The preparation stage involved the crushing and grinding of the raw carcass material to produce a paste with about 40% moisture content. This was fed into a large rectangular hopper capable of holding 20 tonnes. The paste remained in the hopper for about

1 h before being passed to the cooking vessels where it was treated at high temperature for a number of hours. Following separation of solids and liquids, there was a further steam sterilisation phase.

The hopper into which the crushed raw material was fed was constructed of galvanised steel plates. It was watertight up to the filling level and was fitted with a lid, which was fixed in place but was not airtight. It was not lagged or lined. The contents were drawn out by two augurs, one at the bottom and a second near the top, driven by externally mounted electric motors. Apart from the fact that augurs cannot completely empty a vessel, the shape of the hopper and the nature of the contents would ensure that a coating of the raw material paste would stick to the inside surfaces. The hopper was not heated, but during production, the area where the hopper was located was quite warm. This would lead to drying and caking of any material stuck to the inner faces of the hopper sides. The hopper was not designed for manual cleaning or inspection of the interior. As a result, a layer of material could have built up on the inside, over the 11 years it was in service.

Production was to a considerable extent automated and was continuous except at weekends. Work stopped at mid-afternoon on Saturday. The hopper contents were allowed to run down so that it stood nominally empty as the last material was processed. At the appropriate time, all plant except the effluent treatment plant was automatically shut down by a PLC. All electrical circuits were switched off except for the control room and an independent supply to the effluent treatment plant. There was no 'live' electricity in the vicinity of the raw material hopper.

Workers entering the factory at 08.00 a.m. on the following Monday found the hopper glowing red-hot. Management and emergency services were alerted and the fire was confined to the immediate vicinity of the hopper. The outbreak had clearly originated within the hopper interior. The two motors were relatively undamaged and had obviously not been involved. The insulation had been melted on all electrical cables running within several feet of the hopper. The outside of the main body of the unit was discoloured up to the filling line, while the cover section was relatively intact. The steel-panelled roof of the building was distorted above the hopper and GRP panels in the roof had been charred, but the damage had not spread beyond the immediate area above the hopper.

Because the steel sides would conduct ambient heat to the contents, residues of raw material would tend to dry and cake on the inside of whatever proportion of the hopper was empty while the rest of the plant was in production. The greatest danger of this occurrence was approaching closedown at the weekends. Even with periodic rinsing, it is likely that residues would build up over time, particularly in the corners. When it reached a suitable moisture content, such material would become liable to spontaneous heating. Because of its high content of abbatoir waste, the material would be relatively rich in the highly unsaturated arachidonic acid. If, for example, part of the accumulated material fell off, exposing relatively dry layers to the atmosphere, spontaneous combustion would be very likely to occur. Once started, burning would slowly spread to all the remaining material, because of its fat content. The appearance of the outside of the hopper after the fire, suggested that the entire inner surface had at some time been covered with glowing char. This is also suggested by its condition when the fire was discovered.

The final incident described here involved not natural materials, but synthetic drying components used in one of the most fire-prone of working environments, a paint spray booth. The premises consisted of a purpose-built industrial unit, one of a complex of similar units on the site. Construction was of concrete blocks with galvanised iron–polyurethane sandwich roof panels. The roof also incorporated large translucent GRP panels for daylight illumination.

The proprietor carried on a spray-painting business, offering a finishing service to the joinery trade. He had been in the business for about 20 years and had occupied his current premises for 5 years. Clients would deliver part-finished wooden doors, panels, cupboards, etc. to the works for spray painting or lacquering and then collect them for final assembly. Much of the material was for onward delivery to the building industry, so there was considerable pressure of work at all stages.

The spray booths were constructed of sheet steel and polyurethane sandwich panels and were about 2 m deep. Solvent fumes were extracted by fans mounted in large vertical ducts that discharged to the air above the roof. There were no electrical contacts of any kind within the booths or extractor ducts. The fan motors were externally mounted above the booths and controlled by switch panels located outside the spray booths. Electrical connection to the fan motors was by means of armour cables passing above the booths. The compressors, the dehumidifier and their switchgear were also outside the booths.

One of the booths was dedicated to spray painting, while the other was used to apply lacquer undercoat and final coat. Disposable filters were mounted on frames at the back of each booth to absorb spray solids and deflect them from entering the ducts. The filters were changed every 2–3 months, depending on the volume of work. Some over-spray also fell down in a fine rain of droplets, which solidified as they fell and collected as small spheres on the floor. This dust tended to build up at the base of the filter, and particularly in the corners between the base of the filter and the sidewalls of the booth and needed to be removed at regular intervals.

The sprays in use were standard industrial products, containing a variety of flammable solvents. Those listed on surviving containers included xylene (mixed isomers), isobutanol and di-acetone alcohol. The warning signs and safety phrases referred to most of the materials as 'Highly flammable' and 'Harmful'. The product containing di-acetone alcohol, a solvent for cellulose-based varnishes and lacquers, carried the additional warning 'Explosive when mixed with oxidising substances'. The material carrying the latter warning was a wood stain and clearly underwent an exothermic reaction on oxidation (a normal part of the drying process of many paints and varnishes). It was therefore liable to spontaneous heating under suitable conditions.

There were five powder fire extinguishers in the spray room, one outside each of the booths, two on the walls nearby and a fifth at the exit door. There were three further extinguishers nearby. Workers had been shown how to use the extinguishers, but fire drills were not held. The extinguishers were 4.5 kg dry powder, stored pressure type and had been recently serviced. The technical data on a surviving extinguisher indicated that it was suitable for fires likely to occur in the materials used on the premises.

On the morning of the fire, a highly experienced operator was alternating work in both booths, because another man was on sick leave. Having sprayed a batch in the lacquer booth, he would leave it to become touch-dry while spray-painting a batch in the paint booth. He worked non-stop on his own through the lunch break because the job was to be collected in mid-afternoon. He helped a customer to carry out some finished items at about 2 p.m. On returning to the lacquer booth, he noticed smoke rising from a heap of spray dust that had collected on the floor in the corner between the wall of the booth and the lower edge of the filter. He ran to the canteen in the next unit to raise the alarm and returned with a kettle of water. He threw the water on the smouldering dust heap but it had no affect. He then used the nearest powder extinguisher and this temporarily doused the smoke. However, smoke rose up again when the extinguisher was exhausted. He went to get more water, but was handed a second extinguisher by someone who came in from the neighbouring unit. This was a water extinguisher. It was discharged at the smouldering dust, but the result was that the dust burst into flame, which then very quickly flared up along the filter and onto the roof and side panels of the booth. With the fire spreading rapidly, the operator was forced to flee and wait for the Fire Brigade to arrive.

The course of events narrated by the worker indicated that the fire was caused by spontaneous ignition of spray waste. On examination, it was noted that the charred material was evenly heated through, with no unburned waste in contact with the floor or wall. This supported the claim that the booth had been clean before spraying started on the day of the fire, as accumulated old waste would have already oxidised and cooled. With no solvent present to become involved in the initial heating, it would have shown as less burned layers under the charred fresh material. Because of the unusual rate of production on that day, a heap of fresh lacquer globules, still oxidising at the surface and producing heat, had built up in the corner of the booth. This is a typical precursor of many sequences that result in spontaneous ignition. A slower rate of work would have resulted in less waste build-up and probably fire would not have occurred. Quick-drying wood stains contain solvents (such as di-acetone alcohol) that are so easily oxidised that they present an explosive hazard in the presence of oxidising agents. The desired reaction is a slower combination with atmospheric oxygen, which results in quick drying of the stain or lacquer, with production of heat. This presents no problem when the lacquer is spread in a thin layer on the target surface, as the heat readily dissipates. However, the heap of waste spray globules left after the fire was about 100 mm deep in the apex of the corner, adequate to provide the insulation necessary to prevent heat dissipation, while the spherical shape of the droplets was optimal for oxygen penetration through the heap. As the temperature rose, the rate of the exothermic reaction would increase, driving the temperature up to the point where the heap began to smoulder. At this stage, lack of oxygen might have delayed the progression to flaming combustion. The kettle of water had no effect, because it was not enough to cool the heap and may in fact have mainly flowed off without penetrating through the surface. The powder extinguisher worked because it reduced the oxygen availability, but not enough powder was applied to blanket the smouldering material thoroughly. The water extinguisher caused the heap to burst into flame, because the water under pressure disturbed the heap and allowed the hottest particles to come into contact with fresh air. The water may have boiled on contact with the inside of the heap, raising a cloud of steam and burning solvent vapour in an effect similar to pouring water on a fat fire. The fire extinguishers provided in the spray works were of the correct sort for the fire that occurred, and the one powder extinguisher used functioned perfectly and was effective. Had a succession of the dry powder extinguishers been used instead of the water extinguisher, it is probable that the fire could have been contained until the Fire Brigade arrived. As it was, the fire had progressed to such a stage that the Fire Officer who arrived first on the scene decided in the interests of safety to wait for back-up units to arrive before allowing fire fighters into the spray room.

The provision of fire blankets would also have been helpful. Unfortunately, there is little or no training available to workers in industry in the use of extinguishers or other first-aid measures in real fire situations. Untrained personnel who find themselves surrounded by dangerously flammable materials in a real fire situation, as happened in this case, are liable to act on instinct rather than information.

Hazardous environments

As stated in the previous section, spray-paint booths are among the most hazardous of working environments. In addition to the danger, in some circumstances, of spontaneous combustion as described earlier, fires are caused by failures of design and maintenance and by human error.

Many paint spray booths in small business premises are poorly designed, constructed and maintained. A common fault is use of a fan with the motor mounted in the duct, compounded by inadequate screening and cleaning, which allows spray residue to build up on surfaces including fan parts and the inside of ducts. Mechanical damage to the fan motor can be caused by the fan blades becoming unbalanced as a result of deposition of spray residue, leading to overheating and fire. The National Fire Protection Association (USA) in its *Fire Protection Handbook 1976* [1] states 'Even motors of the enclosed, explosion-proof type are not suitable for use inside spray-booth ducts because accumulation of residue on the outside of the motor will interfere with normal cooling of the motor'. The motor should be mounted outside the duct, driving the fan by means of a shaft or enclosed belt. The fan housing and blades should be of non-sparking materials. It is further recommended that ventilation of each spray booth should be by a separate exhaust duct leading by the shortest route to the outside of the building. Inspection ports should be installed to facilitate cleaning of the duct and fan and cleaning should be regular and adequate to avoid build-up of solid residues. However, it is not unusual to find, in the aftermath of a spray booth fire, that the duct has not been provided with access for cleaning and the fan has not been serviced, cleaned, or even inspected since installation, often several years before the fire.

In some cases, there is no interlock between fan and spray operation, so that it is possible to start spraying without commencing ventilation. In one such instance, the operator, realising that he had forgotten to switch on the fan, stepped outside the booth to remedy the omission. There was an instantaneous eruption of flame from the direction of the extraction duct, which spread to engulf the booth.

Even where best practice is used and the booth and equipment are of good design and well maintained, inadequate fire safety training can lead to accidents. The following accident occurred in a well-designed booth in a small furniture factory, which was electrically safe and where a suitable fan was correctly installed and maintained. Spraying was carried out by an operator using a compressed air-driven lance, attached to air and paint-lines that passed out through the wall of the booth where they were connected to drums of spray lacquer and to a compressed air line. The equipment was electrically bonded (earthed) to prevent static build-up in the lance or the spray droplets.

A fire occurred while an operator was applying cellulose-based base coat to a piece of furniture. He suddenly found himself surrounded by flames, dropped the spray-lance (which cut out as it was designed to do) and ran towards the booth safety doors, which opened outwards and were held closed only by external air pressure. The bruises he received in rushing through the doors appeared to be the only injury he suffered, although he was severely shocked by the episode. There was a short-lived rush of flame from the open doorway and from the hinge sides of both doors when they were opened by the spray operator, but otherwise the spray booth contained the fire.

There was no apparent source of ignition in the structure or fittings of the booth. There was no evidence of malfunctioning of the lighting or fan prior to the fire. It was clear from the burning pattern on the walls that the fan continued to function for a time after the fire started, although the aluminium blades were eventually melted. Apart from a ban on smoking on the premises, it was not possible to smoke and spray at the same time, because the spray operator wore a mask that covered his eyes, nose and mouth. The man had been on his own in the booth when the accident occurred.

According to the proprietor and other personnel present at the time of the fire, the morning had been cold and dry. It was policy to keep the atmosphere in the premises cool and as dry as possible to prevent the wood from imbibing moisture. The spray operator was described as wearing a 'hairy' sweater, probably a synthetic mixture, under his overalls, which were made from a polyester/cotton mixture. Both these types of materials can build up static charges as a result of friction when worn, particularly in cool, dry conditions. Where garments have a surface pile of long fibres, considerable static charges can build up, so that individual fibres stand out from the garment surface. Frequently, as the wearer moves about, static discharges occur between one garment and another, or onto some other object. The discharge causes a spark that is capable of igniting flammable gases or vapours. In this case, it is not unlikely that a discharge occurred from the worker's clothing to the earthed spray-lance. Despite a clear effort to make the spray booth as safe as possible, and instructions to spray operators not to wear nailed boots so as to avoid friction sparks, the management were unaware that static discharge from certain types clothing was capable of igniting flammable vapours. The danger could be avoided by ensuring that personnel operating spray-paint equipment wear cotton external clothing and overalls of cotton or disposable paper. It would also be an advantage to cover the hair with a cotton or disposable paper cap. These materials imbibe and retain atmospheric or body moisture and so become conducting, allowing static charges to drain away instead of building up to the point where a large potential difference develops between the garment and its surroundings, leading to a spark discharge.

Particular hazards also attach to the use of deep fat cooking equipment in both domestic and industrial situations. Deep fat frying is carried out at temperatures close to the ignition temperature of the fat, so that there is a necessity to exercise particular care during the operation. Fires are often the result of trivial sparks or momentary inattention, but can develop into serious outbreaks because of the presence of excess grease or other sources of fuel in the vicinity of the original ignition.

A typical sequence of events occurred in a hotel kitchen as preparations for breakfast got under way. A gas-operated deep-fat fryer consisted of a metal cabinet about one metre tall and a working surface of about 400×300 mm. The upper part was occupied by an insulated trough to contain the frying medium (vegetable oil), with a steel mesh screen about 200 mm below the top, to stop food sinking to the bottom of the oil. It would not have stopped small-size particles such as breadcrumbs from falling into the oil. The food was fried in steel wire baskets, placed above the fixed mesh screen. There was no lid fitted to the oil trough. The oil was heated by a gas burner located in the cabinet below the trough. Ignition was by pilot flame, left burning continuously. There was a 'flame failure' safety device to monitor the pilot flame. This operated valves to cut off the gas supply to both the pilot light and the main burner in the event of extinguishment of the pilot flame. The gas flow to the burner flame was controlled by a rotary switch, marked from 0 to 8, located on the front of the burner assembly in the cabinet. There was no temperature regulating device other than the gas flow control and no thermostatic cut-off in the event of the oil overheating. It was necessary to open the cabinet door to operate the gas flow control.

The oil in the fryer had been in normal use for four days prior to the fire, just over half its expected use period. The fresh oil had been taken from a container already opened and partially used. It had not shown any unusual heating behaviour. The parameters shown on a new drum of oil in the kitchen included:

Smoke point	243°C
Flash point	338°C
Ideal for chips	182–188°C

It was the duty of the night porter to switch on the deep-fat fryer in the morning, in preparation for the cooking of breakfast. On the morning in question, he went to the kitchen at 06.45 a.m. and turned the gas regulator on the fryer to the highest point, 8 on the dial, which was his usual practice. He then left the kitchen and went to Reception to wait for the arrival of the female staff member who served the breakfasts. He admitted her, chatted for a few moments, then returned to the kitchen to turn down the gas on the fryer. He believed that it had been turned to the highest heat for about twenty minutes, which again was the normal practice. He found the oil in the fryer smoking heavily when he returned to the kitchen. He turned down the gas and assumed that the oil would cool down. However, he had just turned away from the appliance when the oil burst into flames. He tried to deal with it by placing wet cloths over the trough, but was unable to control the fire. He returned to Reception and raised the alarm.

As a general principle, deep-fat fryers and pans should never be left unattended when switched on, even for a few minutes. Users tend, as happened in this case, to start operations by turning the heat full on in order to raise the oil temperature rapidly to cooking level. Most do not understand that because the appliance is insulated, the oil will take some minutes at least to cool down. There was no evidence of defect in the deep-fat fryer. The chef was aware of its lack of an emergency cut-out (not an unusual omission) and apparently used it without problems but with due caution. The night porter however was not apparently aware of the danger of leaving the appliance unattended.

The principle of deep-frying in a bath of hot oil or fat is that the food can be rapidly cooked through without any corresponding change occurring to the oil. This is because the food generally has a much lower ignition temperature than the oil. Some foods may be covered with particles such as breadcrumbs or other cereal-based coatings with a still lower ignition temperature, so that the pieces become browned (i.e. partially pyrolised) on the outside, again without any corresponding change occurring to the oil. The oil can therefore be used over again many times. However, the addition of substances (other fats, juices) from food as it cooks may cause lowering of the boiling point of the cooking oil, leading to smoking at lower temperatures. Breadcrumbs and similar particles with lower ignition temperatures than the oil itself are deposited and soon become charred. In this case, the smoke would have been contributed both by the oil, as it neared boiling, and also by the contaminants, particularly breadcrumbs and other carbohydrate material, now charring in the hot oil. Although the porter lowered the gas flow, the oil, in its insulated container, would not instantly cool down and the particles contained in it would continue to char. His entrance into the kitchen and his movements in the vicinity of the deep-fat fryer probably resulted in an increased air-flow to the surface of the hot oil, providing sufficient oxygen to allow some of the charring carbohydrate particles to ignite, in turn igniting the oil.

The problem of leaving deep fat fryers to heat unattended is even greater where the appliance is heated electrically. It is possible for the oil in deep fat fryers to ignite even if the thermostat is fully operational. This is because the heating element must always be at a much higher temperature than the oil in order to heat the appliance up quickly. The thermostat switches off the power when a pre-set temperature is reached, but the element remains at high temperature and the oil temperature continues to rise for a short time and the oil vaporises at the surface.

Cooking appliances may not be adequately maintained. As with spray booths, the extraction equipment in industrial kitchens is often poorly designed, with long and tortuous duct pathways to the open air. Fat condenses and is deposited in the duct, particularly at bends, where it can be ignited by sparks drawn up from otherwise trivial flare-ups at the cooking hob. Ducting is almost never supplied with adequate access for cleaning and may be partially constructed of unsuitable material (e.g. aluminium) and without adequate separation from wood or other combustible materials. As a result, what begins as a trivial incident may result in major structural damage.

Domestic fat fires, typically in chip pans, usually result from allowing the pan to get too hot, or leaving it unattended. Alcohol may be a feature, where a fat pan left to heat on a gas or electric ring is then forgotten, or the user falls asleep. Unfortunately, the outcome can too easily result in loss of life, particularly where there are children or older people in the household. If the fire spreads to kitchen fittings, production of toxic smoke may trap occupants on upper floors. The provision of fire blankets in domestic kitchens needs to be encouraged, as well as the installation of smoke alarms in the circulation areas of domestic dwellings. The use of properly designed domestic deep fat fryers should be more strongly promoted and parents should be made aware of the danger of allowing children to attempt to prepare chips without adult supervision.

Conclusion

Investigation of accidental fires and compilation of data on the causes is carried out by Fire Services and the Insurance industry in many developed countries, but the statistics, even if made available, are not always in a form accessible to the general public. The investigator may report breaches of regulations, ignorance of standards, or plain lack of common sense, but unless people can be educated to be aware of their personal responsibility, accidental fire will continue to damage lives and property in the home and workplace.

References

1 *The Fire Protection Handbook* (1976) National Fire Protection Association, PO box 9101, Qunicy, MA 02269-9101, USA.
2 D. Drysdale (1985) *Introduction to Fire dynamics*, first edition, John Wiley and Son, New York.
3 BSI PD 2777: 1994, British Standards Institution, HMSO.
4 Dangerous Substances Act (1972) Government Publications office, Dublin.
5 BS 5438: 1989, British Standards Institution, HMSO.
6 Industrial Research and Standards (Section 44) (1991) Petroleum Coke and other Solid Fuels Order, Government Publications Office, Dublin.
7 Barbour Index Health and Safety Professional (1989) 7 Welding and other hot work processes.

8   Factory Mutual Engineering Corporation booklet (1997) *A Pocket Guide to Hot Work Loss Prevention*.

9   H.P. Rothbaum (1963) *Journal of Applied Chemistry* 13: 291–302.

10 J.A. Currie and G.N. Festenstein (1971) *Journal of the Science Food and Agriculture* 22: 223.

11 P.C. Bowes and B. Langdorf (1968) Spontaneous ignition of oil-soaked lagging, *Chemical and Process Engineering* May.

3

Electricity and fire

John D. Twibell

Introduction

In almost every significant fire in a building the fire will eventually reach and burn through energised electrical cables and produce short circuits in the wiring. If this happens in what appears to be the seat of fire region the investigator will be faced with the question as to whether the signs of 'electrical activity' in the burnt wiring are the cause or a consequence of the fire. Unfortunately, the relationship between electricity and fire is often misunderstood by many investigators whose reports on fire form the basis of national statistics in the UK.

The National Fire Statistics in the UK compiled by the Home Office appear to show that about 25% of fires are 'electrical' in origin, but this hides a multitude of sins. First, the statistics are compiled from fire report forms that ask for the 'supposed cause' to be entered. This may be completed by a fire officer who has had no training in the investigation of fire and to whom the finding of a burnt piece of electrical cable showing short circuit arc marks is clear proof of electrical causation, whereas it is almost certainly the result of a fire having burnt through the energised cable. Second, a number of fires are reported as 'electrical' if they occur through misuse of an electrical item. Thus a fire caused by the ignition of clothes placed too close to an electric radiant heater may be categorised as electrical. Similarly, a chip pan fire may be reported as electrical if it occurs on an electric cooker.

The extent of the misreporting can be seen from a recent survey [1] of fires fully investigated by the Forensic Science Service (FSS) Huntingdon Laboratory over a five-year period (1990–94 included). This survey showed that only 2% of these fires were actually caused by electrical faults. It could be argued that as the FSS was only called in to investigate fatal or suspicious fires, these fires were biased away from accidental causes. However, during the survey period it was in police force's Standing Orders to ask the laboratory to investigate all fatal fires. It might be expected that if electricity was a major cause of fire this would show through in fatal fire statistics, but electrical causes were found to account for only 6% of the fatal fires examined.

The widespread misreporting of fires as 'electrical' can have serious effects during arson trials. Here it is often put to the prosecution investigator by defence lawyers that the fire was actually caused by an electrical fault. Thus the investigator has not carried out the scene examination properly by having missed the true cause. If the defence can show that the investigator has missed or ignored some electrical item in the supposed fire seat area this will be used to discredit his findings. It is therefore imperative that even if the cause of the fire is clearly deliberate, investigators must spend further time at the scene in examining electrical items to exclude any such later challenges.

Britain is not alone in having defective national fire statistics wrongly biased towards electrical fires. In the USA and Canada similar complaints and calls for better understanding and training have been made by specialist electrical fire investigators such as Ettling, Béland [2,3] and others.

Purpose and reader requirements

The purpose of this chapter is to try to explain and educate the knowledgeable investigator in the problems posed by the interrelations between electricity and fire. Its aim is to explain basic electrical installations and try to explain various potential fire causing faults, their effects and the possibilities of their occurrence in a properly protected circuit. The information is given in a relatively simplistic manner and may therefore appear to be superficial to the trained electrical engineer.

A basic knowledge of electricity has to be assumed. It is expected that the reader understands electrical units such as volts, amps, ohms and watts and their interrelations. It is also expected that the reader has limited knowledge of the electricity supply system, and its distribution in a domestic installation. Although this chapter deals mainly with UK AC mains supply, the term *resistance* is largely used rather than the more correct *impedance*.

IEE Wiring Regulations

In the UK almost all aspects of electricity distribution in buildings are covered by the Institution of Electrical Engineers (IEE) 'Requirements for Electrical Installations' or 'Wiring Regulations', as they are also known. The 16th edition (1991), was incorporated by the British Standards Institute as BS 7671 in 1992. At the time of writing the latest edition is a full revision of the 16th edition designated as BS 7671: 2001 [4] which came into effect in January 2002. The Regulations are numbered under various headings, and run to approaching 300 pages, including tables and diagrams.

The Regulations apply to wiring and installations in almost every type of building or scene likely to be encountered but there are notable exceptions. Thus, they do not apply to the actual electricity supply company's connections to the premises (which are 'suppliers works' as defined by The Electricity Supply Regulations 1988). Also excluded from the scope of the Regulations are railway traction and signalling, motor vehicles (other than mains wiring in caravans, campervans, boats, etc.), and installations on ships, aircraft or in mines, etc.

The Regulations are designed to protect persons, property, livestock, etc. from electric shock, fire, burns and injury. They are essentially Codes of Practice and are non-statutory regulations, but they may be used in court as evidence to claim compliance with statutory requirements. In general the Wiring Regulations are not retrospective and if a building is wired to the regulations pertaining at that time there is no immediate requirement to change the installation if the regulations change.

The Wiring Regulations themselves are somewhat dry and not easy to read or follow through, but the IEE also produce a number of easy to read explanatory booklets. These include a *Guide to the Wiring Regulations* and a series of *Guidance Notes*, number 4 of which is '*Protection Against Fire*' [5].

Electricity at Work Regulations 1989

It is a legal requirement that these Regulations are complied with in any workplace and they have particular relevance at fire scenes where there is a danger that the electricity supply may not have been isolated. The main points concerning this Statutory Instrument and their implications for scene or laboratory examinations are discussed in a later section of this chapter.

Terminology

Over a series of Wiring Regulation editions and updates, terminology changes have occurred which may prove confusing. Thus what was once known as *the earthing continuity conductor* (ECC), was later known as the *circuit protective conductor* (CPC) and *earth leakage currents* have now become *protective conductor currents*. The terms *earthing conductor* and *earth leakage current* are preferred in this text as being more explanatory to the non-specialist investigator. As much of the text refers to single-phase circuits the term '*Live*' is largely used, as being more explicit than '*Phase*'.

Mains voltages

The nominal mains voltage in Great Britain had been standardised at a nominal 240/415 V rms (i.e. 240 V for single phase, 415 V for three phase) over a long period. The supply to consumers was allowed a tolerance of $+/-6\%$. At that time much of Europe had supplies of nominally 220 V ($+/-10\%$). On 1 January 1995 UK Statutory Instrument 1994 No. 3021 – The Electrical Supply (Amendment) (No. 2) Regulations 1994 came into force as part of the progression towards European harmonisation, the nominal voltage across Europe changed to 230/400 V with various regional tolerances. It is intended that all of Europe harmonises to 230 V $+/-10\%$ in 2003 which still allows the UK to supply approximately 240 V and others at 220 V. In accordance with this the mains voltages referred to in this text are 230/400 V. Most persons are unaware that the mains voltage has been nominally reduced to 230 V. Many recently manufactured electrical items are now marked for 230 V operation, in line with the harmonisation changes. This does not mean that they are unsuitable for use on the British mains supply, and in particular that their use might lead to fire.

Electrical circuits, components and protective devices

The distribution system

Most final users in the UK are supplied from an electricity supply substation that incorporates a three-phase transformer. Essentially the output side of the transformer has three windings each of which supplies the three differently phased outputs (Figure 3.1).

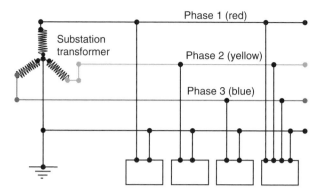

Figure 3.1 Local electricity supply distribution to consumers (e.g. three single-phase domestic consumers and a three-phase supply to a larger building)

The other ends of the three windings are connected together to form the Neutral point of the transformer, which is connected to Earth at the substation. The Neutral connector for the entire distribution system is taken from this point. The three outputs each produce a sinusoidal waveform at the mains frequency of 50 Hz but the three phases are staggered at 120° phase difference apart. Each winding thus produces a sinusoidal output varying from positive to negative with respect to the earthed Neutral connector or effectively producing a supply to the consumer of nominally 230 V rms with respect to the Neutral. Due to the phase differences between the three windings a sinusoidally varying potential is also generated between each pair of the three windings that is nominally 400 V rms. Substations usually have an automated mechanism for adjusting the output voltage to accommodate voltage drops caused by increased loads, etc. They also include fuses in each phase to protect the transformer from high overloads or faults. Cables take the Neutral and the fused output from each phase into the distribution circuit. Locally, the supply may be by underground or overhead cables. In an attempt to balance the loads, consumers are distributed across the three phases as shown in the diagram. A three-phase supply will be given to small factories or service stations, etc.

Domestic single-phase supplies

In a supply to a single-phase domestic consumer the Phase (Live) is fed through a service fuse (often known as a 'cut-out'). The Neutral conductor is fed to a Neutral block which (depending on the type of installation) may be connected to Earth at the premises or the householder may have a separate Earth. The Live and Neutral are then fed through the meter to the householder's distribution circuitry (Figure 3.2). In a three-phase supply each phase is fed through a separate cut-out fuse and (with the unfused neutral) thence through the three-phase meter to the distribution boxes. The installation from the incoming service cable through the cut-out and meter is the property of the supply company. The installation after the meter is the property and responsibility of the householder.

The supply from the meter may feed one or more distribution fuseboxes, modern versions of which are often referred to as 'consumer units'. The consumer unit houses fuses or MCBs

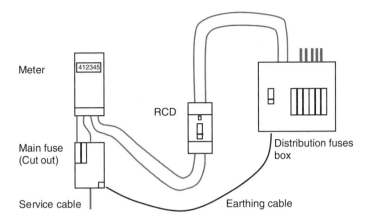

Meter

RCD

Main fuse
(Cut out)

Distribution fuses
box

Service cable

Earthing cable

Figure 3.2 Typical domestic supply installation (1980s). Earlier installations may have no RCD or be via an ELCB to the consumer unit, later ones are likely to have a split load consumer unit with an integral RCD supplying selected circuits (see text)

(Miniature Circuit Breakers) feeding the various circuits in the installation. In a domestic situation these may be the final fuses/ MCBs in lighting, heating or cooker circuits. For ring main socket circuits the fuses/MCBs will be rated at 30 or 32 A and the final circuit fuses will be the 13 A or smaller rating cartridge fuse within the three-pin plug on each appliance lead.

Lighting circuits and supplies to fixed high load equipment such as cookers, water heaters and showers are made by means of radial circuits where a single cable runs out from the circuit fuse in the fusebox to the load. In these circuits the cable must be adequate to carry the full current of the load. In the case of lighting circuits there may be one for upstairs and one for downstairs in a dwelling and a single cable usually runs to sequentially supply several lighting points in turn. Socket outlets in the UK are supplied mainly by ring main circuits. These circuits are designed to supply up to 30–32 A but two runs of cable go out from the fuse in the fusebox to supply each end of a string of sockets connected together to form a ring, with the Live, Neutral and Earth conductors each looped around the whole circuit (Figure 3.3). Thus, the current drawn by a load connected to any socket in the ring is actually being shared by both cables, which allows slightly less substantial cable to be used. Secondary circuits known as 'spurs' may be connected via single cables into the ring but the current drawn down the spur should be limited by appropriate fusing.

Within the distribution circuitry the wiring should be colour coded with the convention of red for Live, black for Neutral, and green and yellow striped (green/yellow) or bare copper for Earth. Blue and yellow are often used to denote switched Live 'strappers' in a two-way lighting circuit. Appliance flexes use the convention brown for Live, blue for Neutral and green/yellow for Earth. Use of a solid green colour as an identifier has not been allowed in recent editions of the Wiring Regulations but it was previously used as the earthing conductor identifier at least as far as the 13th edition. Hence solid green insulated conductors can sometimes be encountered in old installations. Occasionally, flexes can be found on old appliances which use the previous flex colours red, black and green for Live, Neutral and Earth, respectively.

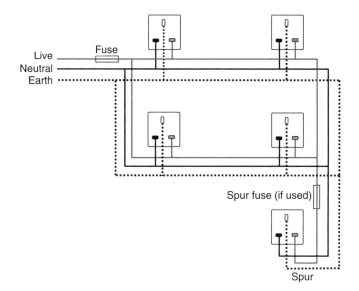

Figure 3.3 Typical ring main circuit with fused spur socket

Three-phase supplies

Most factories, industrial units and public buildings receive a three-phase supply. In many situations it may not be immediately obvious to the occupants that the building has a three-phase supply because the phases are separated into three separate single-phase 230 V supplies. Thus the building will have ordinary 13 A ring main sockets exactly as in the domestic situation, and unless a full three-phase supply is required to run an industrial motor, the average worker may be unaware of the difference. Examination of the distribution system would, however, reveal the three-phase incoming supply, with its three fuses and three-phase meter and the distribution switchboards and fuse boards usually contain all three phases colour coded (red, yellow or blue with black for Neutral) within each box. Clearly, the difference is that if a fault occurs between phases there is a larger voltage difference available (400 V).

Because most three-phase supplies are broken down into single-phase circuits for final use, the distribution wiring should not be that much more difficult to understand than in a domestic installation. It will, however, contain many more circuits and depending upon the size of the installation may contain a few or many individual distribution or isolator boxes. The investigator must be capable of understanding what these boxes do in terms of isolating supplies or distributing the supply to other areas through successive tiers of fusing, etc. He or she must be capable of deducing which parts of the distribution contain a three-phase supply and where the phases are separated to the final circuits. Part of an investigator's training should involve examining unburnt three-phase supply boxes to gain familiarity with such systems.

Where all three phases are present the cable insulations should be colour coded to denote the phase. Once separated into individual single-phase circuits the wiring colour-coding convention reverts to red for Live. A typical three-phase distribution fuse box is likely to contain a master isolator switch that switches all three phases from the same toggle or switch handle. Generally the fuses or MCBs are arranged in three rows or columns (one per phase), but sometimes they may be arranged sequentially in the same bank. Where the fuses/MCBs are used to supply single-phase circuits the Live phase output cables from each phase are likely to be coded red, particularly, if they go finally to an area where confusion between phases is unlikely to occur. Thus, the trunking from that box is likely to contain numerous red cables, which are actually on differing phases. Sometimes electricians may add a colour-code tag of red, yellow or blue tape or similar to a length of red cable to denote its phase, but this is not usual. If a three-phase supply is taken from this box to a three-phase motor or similar, it will require a feed from a fuse on each phase bank and these cables should be colour coded red, yellow or blue to denote the particular phase. Under recent proposals for further European harmonisation, the UK may have to change its phase identifier and live and neutral colours for new installations from 2006.

The lighting load of many larger buildings is usually spread across all three phases. This is particularly important on factory floors or workshops where fluorescent light fittings are usually split into multiples of banks of three, separately fed from the three-phases. This is to fill in any undue flicker in the system and to try to avoid stroboscopic effects, which might otherwise suggest to workers that rotating machinery is stationary.

A particular feature of three-phase supplies is that if the loads on all three phases are balanced (i.e. the same), the current effectively flows between the three-phase terminals (via the loads and the local interconnections in the neutral conductor), and no current would flow in the Neutral conductor back to the substation transformer. If the load was perfectly balanced, as in a three-phase motor, there is thus no need for the Neutral conductor. If the Neutral became disconnected, the motor would still run and little or no damage or effect would be noticed. In the case of a three-phase supply split between individual consumers, the effect would be much different if the Neutral connection was lost, as described later under 'Loss of Neutral link'.

Cables and their current carrying capacities

Conductor sizes are quoted by cross-sectional area (mm^2). Whenever a current flows through a conductor, some heat will be produced as the current overcomes the internal resistance of the conductor. For a given length of cable, the power dissipated in the cables is the product of the square of the current flow through the conductor and the cable's resistance (I^2R). The actual amount of heat produced in the cable is a function of the time (in seconds) over which the current flows such that the heat produced (joules) $= I^2Rt$. There is a current limit for a particular conductor size beyond which the heating effect can cause damage to the cable insulation. Nowadays conductors are rated by their cross-sectional area, and at 30°C in free air. The current rating is the maximum current which the conductor can take before heating to a temperature that is likely to damage the insulation, which for ordinary PVC cables is taken to be 70°C. The thermal environment of the cable is also important as, if the heat loss from the cable is restricted its final equilibrium temperature will be higher. Appropriate factors are therefore applied in the regulations to derate the current carrying capacity of the cable to allow for elevated temperatures; multiple cables within the same enclosure; and the presence of thermal insulation.

Cable construction

In fixed installation cables (i.e. for domestic or industrial fixed wiring) each current carry-ing 'core' is usually a single conductor strand up to the 2.5 mm^2 (1.78 mm diameter) size.

Above this size conductors usually have multiple strands as single strands would be difficult to bend and there is the danger of the conductor fatiguing and breaking if it is bent too often. Flexible cables ('flexes') for connecting mains supplies from the outlet socket to the appliance need to be much more flexible as they are likely to be moved around repeatedly and flexed from side to side. The conductor cores, therefore, are of multistrand annealed copper construction.

There have long been problems in referring to different types of cable insulation and anomalies within the earlier classifications. BS 7671: 2001 [4] now refers to only two types, namely *thermoplastic* and *thermosetting*. PVC cables soften and melt when heated and can be moulded, hence they are designated as *thermoplastic*. Most rubbers and many other materials are cross-linked and cannot melt and be remoulded when heated. They eventually pyrolyse as solids and are referred to as *thermosetting*.

The type of insulation used dictates the final end-use of the cable, particularly in terms of its thermal environment. Thus ordinary PVC insulations are used where it is expected that the cable temperature will not exceed 70°C. Some thermosetting compositions may operate at up to 90°C, and mineral insulated cables can operate at even higher temperatures. Ordinary PVC installation cables and flexes can become very stiff and even brittle at low temperatures and should not be used for outdoor applications in the winter. Specially designed 'Arctic' cables use a plasticised PVC composition which remains soft at low temperatures (ranging from −20 to +70°C).

Multiple core cables (i.e. containing several conductors) are double insulated, containing the individual insulated conductor cores within an outer sheath of similar insulation. The main exception is Twin and Earth (T&E) fixed installation cable which has a bare copper Earth conductor sandwiched between the insulated Live (red) and Neutral (black) conductors within the outer insulating sheath.

Fixed installation cables

The ratings given in the Wiring Regulations allow a respectably wide margin of overload. Experiments carried out by the author show that fixed installation cables can generally be overloaded by a factor of 2.5 or so before the cable runs into serious overheating, but the degree of overload acceptable is modified to a great extent by the cable environment. Proper installation design should ensure that the fixed installation cables are suitably sized for their purpose. Whilst it is sometimes suggested that older PVC wiring may be more susceptible to overload since nowadays users tend to add extra sockets to their installations, in older installations wiring was quite often over-specified and had capacity for the introduction of further loads onto the circuit. Thus unless an excessive load is applied to a particular circuit, coupled with an inappropriate uprating of the circuit fuse, the cable installation will almost certainly cope. (The modernisation of such installations by adding further sockets or spurs does not necessarily add appreciable load as the new sockets are often to accommodate the proliferation of computer and entertainment items with very low power consumption.) On the other hand it may be that more modern installations pose more problems. Modern pricing practices often adopt the minimum specification to keep the job cheap and undercut competitors, in which case the installation is unlikely to cope with an increased load. (It should be noted that BS7450/IEC1059 1991 states that designers should consider the economic life of the cable. Thus if the cable becomes overloaded its life will be shortened.) In dwellings, further problems may be caused by Do It Yourself enthusiasts who lack electrical knowledge.

Flexes and extension leads

Flexes are normally supplied with the appliance and often already have a moulded plug containing the correct size of fuse. Extension leads normally display a maximum current rating and coiled extension leads display warnings that the entire length of the cable should be unrolled prior to use. An unrolled extension lead rated at 13 A fed via a 13 A fuse will behave appropriately on a 3 kW load (13 A). However, if the cable remained wrapped on the drum the heat generated within each layer or adjacent coil cannot easily be dissipated.

Figures 3.4 and 3.5 show a typical cable overloading experiment and the temperature profiles obtained. The experiment follows the differences in temperature achieved in the same length of cable carrying a given load but with different parts of the cable length running in different environments. Sensor 1 shows the temperature rise of the cable in free air, sensor 2 the rise where the cable passes through thermal insulation and sensor 3 where several layers of the cable are wound on a drum. For this particular experiment it can be seen that the cable wound on the drum produces the worst case scenario and the cable reaches its maximum equilibration temperature after about an hour.

Dependent upon the degree of overload a coiled cable will continue to heat beyond the point where the insulation softens and melts or pyrolyses. In extreme cases, the insulation

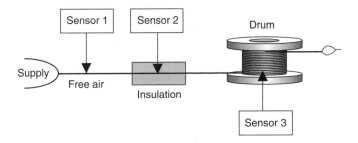

Figure 3.4 Overloading experiment with cable running in free air, with thermal insulation and wound on a drum. Typical experiments use a low-voltage high-current source to feed the twin, or T&E cable which is shorted at the far end

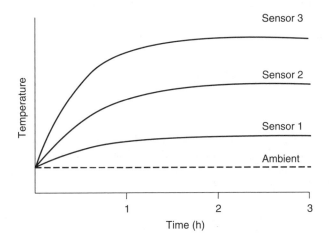

Figure 3.5 Temperature profiles from cable overloading experiment (Figure 3.4)

69

will be destroyed and the conductors will either touch together and short circuit, or an electrical arc may be established. A short circuit will almost certainly blow the fuse or MCB and cut off the current, but an arc may be limited by the propagation medium at a level below that required to blow the fuse. In either case, the short duration arc or sparks ejected during the short circuit or the longer stable arc may ignite the pyrolysis products or plasticiser vapours to produce a flaming fire. Such ignition can only occur if the arc occurs in a location where there is air and flammable products mixed within the flammability range (or explosive limits) of the products. Failing this smouldering ignition of some types of pyrolysed insulation may occur. Dependent upon the circumstances a fire resulting from an overloaded cable drum would be expected to occur half an hour or so after connection.

Rubber covered cables are often thought to represent more of a hazard than PVC as the pyrolysis products are more flammable. However, the plasticisers used with PVC are themselves flammable and (if ignited) will burn vigorously when vapourising from a sufficiently over heated cable.

Fuses and MCBs

At its simplest a fuse is a weak link in the circuit that is intended to overheat and burn out if the circuit becomes overloaded. The fuse blows in a safe environment where it cannot cause a fire or personal injury. The fuse is there to 'clear' the fault and thus to protect the circuit from too high a current flow. It also protects earlier parts of the supply circuit against overload. It should be noted that a fuse is designed to 'blow' and in so doing to open the circuit when an excessive current flows. It does not 'fail' when it operates. A 'failed' fuse would be one which either did not blow when it should have done so, or exploded or operated in some catastrophic manner in the process.

Early fuses were simply short lengths of copper wire held in a suitable housing. Later versions of rewirable fuses are still widely used today in older consumer units and are known as 'semi-enclosed' fuses. The fuse wire element is held in a porcelain carrier against a porcelain or similar heat resisting base unit. Tinned copper wire is normally used as fuse wire, the tin offering some protection against oxidation.

When too high a current flows through the fuse element it heats up towards its melting point. The resistivity of the copper increases with temperature and that of the molten metal is even higher. Thus as the link begins to melt there is a sudden increase in its internal resistance coupled with increased heat generation. As the molten metal begins to fall away and break the circuit an arc is generated which tries to maintain the current flow across the breaking fuse link. The combined effect is to cause the molten metal to splash out in an explosive manner as the circuit opens, often accompanied by an audible bang.

Fuse technology

A simple piece of copper wire has relatively poor blowing characteristics. It can burn out very quickly in a large overload or short circuit situation, but it can take a long while for it to blow under small overloads. If a fuse of this type is operated for a long period at close to its limit it will tend to degenerate and weaken due to oxidation. At some later stage it is likely to blow at a current that it would have easily withstood earlier.

Cartridge fuses offer much improved operational characteristics. At its simplest, the cartridge fuse may be a small ceramic tube with tinned brass or copper end caps linked by a thin piece of fuse wire (Figure 3.6) The tube is filled with silica sand which helps to

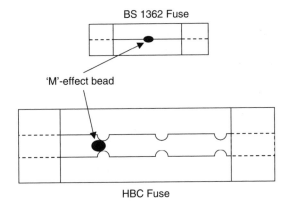

Figure 3.6 Construction of cartridge fuses (see text)

quench the arc by falling into the gap as the link melts. Silver is usually used as the fuse wire as it is more inert and has a lower melting point than copper. A small bead of tin is placed in the centre of the fuse wire to improve its low overcurrent blowing characteristics. This utilises the 'M effect' (Metcalfe effect). At low overcurrents, the heating of the fuse element causes the tin bead to melt and to begin to alloy with the higher melting point metal of the fuse link. The alloy formed has a higher resistance than that of the element and thus increases the localised heating effect. This causes further alloying within the element eventually resulting in a quicker blow than would be the case if no bead was used.

Larger cartridge fuses use one or more elements stamped from a ribbon rather than circular cross section wire. As shown in Figure 3.6 the element contains constrictions, which effectively concentrate the current flow and its heating effect. At very high overloads or in short circuit conditions the fuse will blow at one or more of these constrictions. A small bead of tin is placed on the edge of one of the constrictions in order to utilise the 'M effect' and thus to improve the operating characteristics at low overload. Higher rated cartridge fuses often contain several individual ribbon elements.

Cartridge fuses have many advantages over rewirable fuses. They confine the blow to within the cartridge, the sand packing helps to quench the arc by filling the gap with high resistance material and melting to absorb some of the arc energy. The arc usually extinguishes permanently when the current waveform falls through the zero point of the cycle and thus occurs within 0.01 s of the arc commencing. Such fuses have a high breaking capacity for a small volume. Even a plug top fuse should be capable of breaking a fault current of 6 kA. They are quicker and easier to replace and have better and highly reproducible characteristics with close tolerances. The advantage to the forensic investigator is that they also retain a permanent record of the blow, which can be subsequently interrogated (see Laboratory examination of fuses).

A characteristic of fuses which is often quoted is the Joule Integral or the I^2t. This is the relationship between the fault current, I, and the time, t, that the fault current takes to blow the fuse. Effectively I^2t is a constant for a particular type and size of fuse at high overload currents. This can be used to calculate the energy dissipated in the wiring of the circuit during fault events, if the resistance of the wiring can be estimated. The number of joules of heat liberated in the cable will effectively be Watts × time (s). Hence the heat liberated in the wiring during the time taken for the fuse to operate is I^2t × the cable resistance. If I^2t is say 1000 and $R = 0.05\,\Omega$ then the energy released in the cable will be only 50 J. Over the length of cable concerned this is unlikely to produce significant effect.

Current ratings

The current rating given to a particular fuse is the maximum current that the fuse will carry indefinitely. It is not the current that will cause the fuse to blow. Dependent upon the design or type of fuse, the fuse will operate on a low multiple (less than a factor of two) of the rating in a period of up to an hour. The characteristics of a given fuse size or type are usually defined or standardised by reference to a time period which is sometimes known as the conventional time. For example, the characteristics of plug top fuses to BS 1362 (of any of the rating range up to 13 A) are based on a conventional time of 30 minutes. This type of fuse should hold a current of 1.6 times its rating for at least 30 minutes but at 1.9 times the rating should blow within 30 minutes.

The greater the current going through the fuse (above its rated value) the faster the fuse will blow. Manufacturers, Standards Institutes and the IEE publish graphs of time–current characteristics of fuses as shown in Figure 3.7. To use these graphs it is first necessary to estimate the 'prospective current' in the circuit. In a short circuit situation this would be the maximum current available from the electricity supplier's transformer/substation allowing for the entire circuit resistance up to the short circuit location. We could also use the figure for the estimated circuit current in an overloaded circuit situation. By following this current up the graph until we hit the fuse curve, we can estimate the time that it would take the fuse to blow under these conditions.

All manufactured fuses will differ slightly from each other in their individual characteristics, and the standards to which they are manufactured allow tolerances or standard gates within which the fuse should operate. Individual manufacturers usually manufacture to a closer

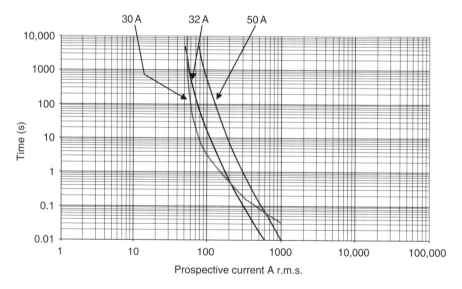

Figure 3.7 Time/current characteristics of a 32 A and a 50 A cartridge fuse (BS88) with that of a 30 A semi-enclosed fuse. (Extrapolated approximation based on data from IEE Wiring Regulations.)

tolerance than the gates and publish tolerances of about $+/- 10\%$ in the current. The characteristic curves drawn for fuses should therefore be recognised as being somewhat broader than their usual thin line depictions. The data used for time–current curves in the IEE Wiring Regulations (on which Figure 3.7 is based) are for the slowest operating times in the various standards.

Discrimination

When there are two or more fuses in the same circuit it is expected that the one with the lowest rating will blow first in a given situation. In any circuit there will be a number of fuses in the line through to the final circuit. Successive tiers of fuses are normally incorporated at each stage that the distribution circuit is broken down into further circuits (Figure 3.8). It is intended that if a fault occurs in a final circuit, the fuse for that circuit should blow and not a fuse further back in the distribution tiering. If a fuse blows further back this will isolate a group of circuits including the fault circuit and will cut off other unaffected circuits. To achieve the desired 'discrimination' fuses in successive circuit divisions or tiers are normally separated by at least a factor of two in rated value. This can be seen in a typical example of a domestic distribution system where the largest rated fuse is the supply company's fuse, (normally 60 or 100 A), the consumer unit fusebox will include various sizes of fuses including the ring main fuses (30 or 32 A) and each of the appliances plugged into the ring main sockets will have a plug-top fuse rated at 13 A or less. In modern domestic and office situations there is a proliferation of extension leads with multiple socket outlets. This means that there may be more than one fuse of the same rating in the final part of the distribution circuit. In such a case discrimination is unlikely to occur. Thus if a series of appliances are plugged in to a multiway extension lead, all will be cut off if a fault in one of the appliances blows the fuse in the extension lead plug, rather than that in the plug of the faulty appliance.

Discrimination works best for fuses of the same type and design criteria which have similar characteristics. In most domestic and many industrial situations the fuses in the successive tiers may be of different types with differing characteristics. Nonetheless the factor of two separation should result in adequate discrimination. Occasionally, however, a fuse can be found to have blown further back in the distribution system leaving the final circuit fuse

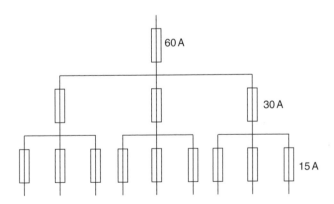

Figure 3.8 Schematic diagram of tiering of fuses in a distribution system. In most domestic or office installations the bottom tier is likely to be the 13 A fuses in the mains plugs supplying the various appliances

where a fault occurred intact. That this can happen can be seen by comparing the characteristic curve for two different types of fuses as shown in Figure 3.7. This shows a comparison between the typical time/current curves for a 30 A rewirable (semi-enclosed) fuse and those of a 32 A and a 50 A cartridge fuse (to BS 88). Here, it can be seen that at low overcurrents the rewirable fuse will blow as might be expected, but for large prospective currents the curve of the 30 A rewirable fuse crosses even that of the 50 A cartridge fuse. Thus if the two fuses were in series in the same circuit, at massive short circuit currents approaching 1000 A, the 50 A cartridge fuse would operate faster than the rewirable fuse. This is a reflection on the fact that the cartridge fuse has better blowing characteristics and will operate faster under severe short circuit conditions. Similarly it is occasionally found that a distribution board fuse will blow in preference to the 13 A plug-top fuse if a short occurs in an appliance or its flex. In situations where the expected discrimination does not occur it may be possible (with cartridge fuses) to show that the lower rated fuse was under extreme stress and about to blow at the moment that the higher rated fuse blew and opened the circuit. Examination as described later may reveal signs of '*short circuit survival*'.

Thermal rupturing of fuses

If a cartridge fuse is subjected to a high enough temperature for long enough, as in a fire situation, the fuse element is eventually likely to break due to thermal operation of the M-effect bead. In a series of experiments it was found that heating plug top (BS 1362) fuses to 450°C for half-an-hour resulted in some thinning of the wire close to the M-effect bead, but the fuse remained intact. Thermal ruptures tended to occur within about 20 min at over 700°C and in about 10 min at 800°C [6].

Miniature circuit breakers

These are electromechanical devices, which serve the same purpose as a fuse. They contain a resettable mechanical latching switch, which operates to break the circuit. They offer the advantage to the householder that they can be easily reset to re-energise the circuit once the fault has been cleared and they can also be switched off manually to isolate the circuit. They offer the disadvantage to the investigator that they do not retain a record of the circumstances, which caused them to operate. Modern units are of a modular construction allowing them to be mounted on busbar systems in distribution boxes or consumer units.

The MCB has a dual mode of operation, combining thermal and magnetic effects into the tripping mechanism. At low overcurrents, operation is by a thermal sensor in which the current passing heats a bimetal strip which, if the rise in temperature is sufficient, trips the mechanism. At very high overloads, the magnetic effect generated in a short winding causes an almost instantaneous tripping effect. The characteristic curves of a series of MCBs can be seen in Figure 3.9. The basic curve is more complicated than that of a fuse, but it is simply the combination of a thermal operation curve (similar to that of a fuse) with the steeper magnetic operation curve, which takes over at very high overcurrents.

As MCBs have a thermal mode of operation their sensitivity is susceptible to temperature variations. They may also be tripped by excessive heat in a fire situation. Thus, if a metal distribution box becomes overheated during the fire it is conceivable that the MCBs may actually trip out even though the supply may have been disconnected. Temperatures sustained over periods of many minutes to hours might be sufficient to operate the trips without

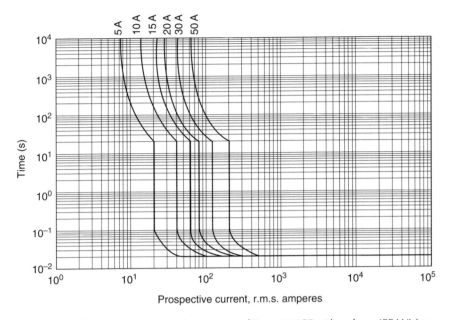

Figure 3.9 Time/current curves for a series of Type 1 MCBs (data from IEE Wiring Regulations)

causing damage to their plastic cases. They are thus a lot more susceptible to thermal effects than are fuses. A related problem lies in the fact that the trip will become much more sensitive to low overcurrents if it becomes heated during the development of a fire.

The finding that an MCB is in the off position after a fire may thus be due to one of three causes. It may have tripped, it may have operated through elevated temperature, or it may have been switched off (before or after) the fire.

Various classification systems have been used with MCBs depending upon their tripping characteristics, but these differences are beyond the scope of this chapter.

Residual current devices

The primary role of these devices is to protect against stray currents leaking from the circuit and thus they are usually thought of as protecting against electric shock. They are designed to open the circuit if a leakage occurs. The earliest type was the voltage operated Earth Leakage Circuit Breaker (ELCB), Figure 3.10. This device was connected within the earthing conductor close to the earthing point such that any current flowing to earth down this conductor could be monitored. Fault current flowing through a resistor in the device produced a voltage which if sufficiently high operated the trip and disconnected the supply. This type of device worked well with leakage currents flowing down the earthing conductor but if the fault current flowed to some other stray earth, the device would not trip.

ELCBs have been largely replaced but may occasionally be encountered in older installations. They were succeeded by the RCD (Residual Current Device) formerly referred to as the RCCB (Residual Current Circuit Breaker). This device monitors the current flowing in

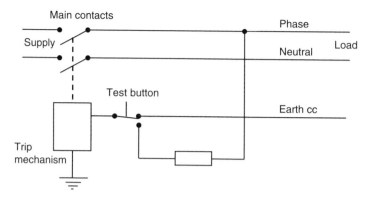

Figure 3.10 Schematic diagram of voltage operated ELCB

both halves of the overall circuit. The Live and Neutral conductors of the monitored circuit or tier of circuits are fed through the device (Figure 3.11). Effectively the Live and Neutral are each fed through the opposing sides of a transformer so that their magnetic effects cancel. In a properly insulated circuit with no stray losses the current going out down the Live conductor should be the same as that coming back down the Neutral. The transformer incorporates a sensor coil in which a voltage is generated if there is net electromagnetic effect. Thus, any imbalance produces a proportional effect in the sensor coil and if this exceeds the trip level the device trips and discontinues the supply. Most devices have a nominal trip current of 30 mA (i.e. an imbalance between Live and Neutral of 30 mA or more). This trip level is based on considerations of the current that the average human body can withstand over very short periods without causing death by ventricular fibrillation.

The RCD offers two distinct advantages over the voltage operated ELCB. First, as it operates only on the difference in current flows, it senses a leakage current wherever that current flows to, whether this be down the earthing conductor or to some stray earth such as a water pipe, radiator or a damp concrete floor. Thus stray current through a person would be detected, but it should also detect a stray current path through some damp building fabric, which might result in a fire. Second if a fault occurs between Neutral and Earth (which may be a few volts apart), the RCD will operate. This should largely remove the possibility that a Neutral Earth fault may cause a fire.

Certain circuits and appliances, such as those incorporating water heaters, tend to have higher leakage currents than other circuits and a 30 mA trip in these circuits may be prone to nuisance tripping. In the late 1970s, RCDs were seen as a large step forward in electrical safety and it became common practice to feed the metered supply via a 30 mA RCD to the consumer units. At this location in the circuit the device (if tripped) will disconnect the whole supply. It has since been recognised that more danger may be created in discontinuing a consumer's lighting circuit at night for a small leakage fault in say a portable appliance, and that circuits supplying freezers or alarm systems ought also to be maintained. Consequently, the approach in the late 1990s has been towards installing split load consumer units where part of the supply (to ring mains socket circuits, etc.) goes via an RCD incorporated within the consumer unit, but lighting and other circuits can be fed via an MCB.

Devices with higher trip ratings and with short time delays have also become available and may be incorporated into the appropriate circuits. British Standards for RCDs recognise

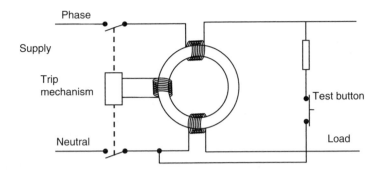

Figure 3.11   Schematic diagram of RCD also known as RCCB

two types, the general purpose (G) type and the selective (S) type with delayed operation. Typical versions available are 100 mA (G), 100 mA (S) and 300 mA (S). The discrimination between devices is such that a 100 mA (S) type of RCD could be used in front of a 30 mA (G) type and will produce adequate back-up and discrimination.

Many items in the home or office now contain semiconductors, which modify the mains AC waveform and can lead to rectified AC (pulsed DC) currents in any earth leakage current. Further, developments in RCD technology allow for this and two types of RCDs differing in their fault current sensitivity are now produced. A new international standard (IEC 1008) classifies those expected to operate in general AC type waveforms as type AC, whilst those which are expected to cope with both AC and pulsed DC are classified as type A.

In recent years RCBOs (Residual Current Breaker with Overcurrent operation) have been developed which incorporate the functions of an MCB and RCD within the same housing, and are mainly of the modular design for use in consumer units or distribution boxes. Typically they incorporate a 30 mA (G) RCD trip together with an MCB function which can be chosen from a range of overcurrent ratings 6, 10, 16, 20, 32, 40 A. RCBOs are also classified as to their sensitivity to AC or pulsed DC under a new standard IEC1009.

As RCBOs contain a thermal operation mode in the MCB side, they too are susceptible to changes in ambient temperature and will trip out on no supply if they become heated in a fire. The finding that an RCBO is in the off position after a fire may thus also be due to one of three causes. It may have tripped, it may have operated through elevated temperature, or it may have been switched off (before or after) the fire.

Arc fault protection

Fuses, MCBs and RCDs would clearly protect circuits against overloads or short circuits and many other faults, but at the time of writing there appears to be no device available in the UK which would disconnect the circuit if an in-line resistive fault, or in-line arcing (see later) occurred. Once arcing develops, it produces a wide spectrum of mains borne and radiated radio frequency interference, which can be readily detected in nearby portable radio equipment. Use of a portable radio may help in detecting some defective connections prior to a fire situation developing, but it is not likely to be useful in stages prior to the development of arcing. Arc fault circuit interrupters are under development in the USA, where early versions are available to the public. It may be some time before they are developed to a stage where they have adequate sensitivity to local faults in the installation and yet can screen out mains borne interference from elsewhere.

Faults which might cause fires

There are a number of ways in which electrical faults might cause fires but certain criteria must also be met for this to happen. In the same way that a flaming match cannot directly ignite structural woodwork, the short-term high temperatures achieved in a short circuit or short duration arc cannot ignite bulk materials. If there are already flammable vapours present from an accelerant, or sufficient pyrolysis products are already being produced from associated heating effects, then self-sustaining flaming ignition may result. Instantaneous faults may be able to ignite more finely divided flammable materials such as thick layers of flammable dusts, woodchips, paper or similar debris, if they are present at the fault location. If the fault occurred within trunking or a connection box there may be insufficient air available to support more than transient flaming combustion. Whatever type of fault occurs, there must be sufficient readily flammable material already present, or already being liberated, together with an adequate air supply to allow sustainable combustion to occur.

The main faults that have potential to cause fires are resistive heating at connections and in-line arcing. Fires may also be caused by the breakdown of insulating materials through 'tracking' or by situations leading to the conduction of stray currents through building materials, particularly if these become damp. In some circumstances major system faults elsewhere in the electrical distribution system can cause fires to develop in the installations of other connected consumers.

It should be noted that although overheating connections may be a problem in electrical circuits, connections should be made within some sort of connection box. Thus most ring main connections are made in the terminals of a socket, at a spur connector or at the fuse box, and other circuit joints may be made in junction boxes. The presence of the housing around a defective connection may limit the possibility of the ingress of water or flammable debris etc. If a fire is thought to have originated from such a connection, its location and the degree of damage to it should be commensurate with it having been the fire cause. Many overheating connections burn themselves out and break the electrical circuit without causing a fire.

Short circuit

A short circuit occurs when two conductors of differing potential come together and make electrical contact. Providing that there is a large potential difference between the two conductors, that there is sufficient fault current available and that the resistance within the rest of the circuit is low, a very large current will flow through the contact. As the contact area is usually relatively small, most of the power in the circuit will be generated within the relatively high resistance of the small point of contact. This causes the metal of the contact point to be heated to its melting point and spattered out with the simultaneous generation of an arc. The typical instantaneous effect to the eye is one of the generation of a sudden arc with sparks of incandescent metal ejected from the contact. The event is usually accompanied by an audible bang. The fault current is usually so large that it blows one (or sometimes more) of the fuses in the circuit and the fault current then ceases. In a normal circuit the overall resistance of the wiring in the circuit may well be less than an ohm or so and thus the prospective current available for a Live to Neutral short circuit would be of the order of several hundred amps. In effect this will amount to a power dissipation within the circuit of perhaps 100 kW over the period of the fault. Much of this power will be created within the short circuit contacts. If the event lasted for say half a mains cycle 1/100th second) this would be equivalent to the rapid release of 1000 J in the circuit.

The resistance of the earthing conductor (earth loop impedance) should also be very low such that a Live to Earth short would also blow a fuse in a similar short timescale. Despite the large amount of instantaneous energy likely to be released in a Live to Neutral or Earth short, the fault current will be rapidly broken by a fuse, MCB or RCD, and is unlikely to be sufficient to cause ignition of normal bulk materials. It could ignite tinder-like materials or vapours at the correct concentration.

Cables do not spontaneously short circuit and if they did so they would be likely to blow the circuit fuse. It is necessary for some extraneous metal item to perhaps bridge two connectors, or for a saw, knife or drill to cut into a cable and cause contact (Figure 3.12). Consider the possibility of a short occurring spontaneously at some point along a run of cable. It would be necessary for the insulation to be removed from a sufficient length of cable to expose the conductors and then for some force to bring the conductors together (otherwise they would maintain their relative positions). This could happen during a period of overloading a cable when the insulation is melting from the conductors and the conductors are thermally expanding or during a fire situation when gravity, stress forces or local structural collapse might bring the exposed conductors into contact. Faced with the common scenario of finding signs of electrical activity (short circuit, arc marks, etc.) at a point along a run of burnt cable which is suggested to be the cause of the fire, investigators should ask themselves why they think that a piece of cable should spontaneously short at this location, and if it did how did it actually cause the fire. How did the insulation become sufficiently damaged and how did the conductors come together? Finally what tinder-like materials or vapours were present at that location to become ignited? If there is some reason for a fault having occurred at this position then the situation may be different. If a cable is trapped in a location where it suffers occasional physical damage or flexing then there may be a reason for a conductor to break and succumb to *in-line arcing* (qv), which may culminate in a short circuit. The appearance of the conductors after a short circuit is discussed later.

It is often suggested as an alternative scenario that *arcing* (qv) developed at the site in the cable and that this led to the fire. Again it is difficult to see why this would happen spontaneously. Even if the cable insulation broke down the conductors are too far apart for arcing to occur unless some other factor such as ingress of moisture can play a part. Béland [3] showed that below 350 V it was virtually impossible to start an arc in a cable at room temperature unless the conductors are brought together to momentarily touch and are then separated. He found that it was equally difficult to start an arc in old cables with brittle insulation, but that irrespective of age, arcing readily occurred in cables heated electrically, or by a fire, to around 400°C.

Overloading and overfusing

Sometimes, it is suggested that the cable had been overloaded and that it must have burnt out or shorted at the damaged location. As previously described, extension leads may seriously overheat if they are not unrolled when supplying heavy currents. In the case of

Figure 3.12   Short circuit caused by sawing through energised cable (Photos: APU/LFB)

properly designed fixed installation wiring with the correct fuses or breakers in place, there should be very little risk of a fire caused by overloading. In tests carried out by the author it was found that a cable could be overloaded by a factor approaching three times its rated value in free air, before serious overheating occurred, and it would take many minutes to bring the temperature up to a dangerous level. If a fuse or MCB of the correct rating was in circuit this should operate well within this timescale.

The IEE Regulations effectively state that the fuse rating should be not less than the design current for the circuit and not more than the maximum capacity of the wiring in the installation. In the case of the 2.5 mm² T&E cable used widely in the UK for radial or ring main circuits, the maximum current carrying capacity is 27 A. If used in a radial circuit (single cable supply) the maximum fuse which should be used is 25 A. A current of about 80 A (about 3× overload) will take a period of several minutes to heat up 2.5 mm² cable to excessive temperatures and start to melt the insulation. Reference to fuse characteristics graphs (IEE Wiring Regulations) shows that a 25 A fuse to BS 88 should operate in about 20 s at this current.

If a 2.5 mm² cable is overfused at say 32 A (BS 88) (or is in a ring main where one side of the circuit has a defective connection), overloading at 80 A should blow the fuse in about 60 s. Thus, it can be seen that overfusing will only be a serious problem if the circuit is heavily overfused.

As previously shown, if a cable is overloaded the heating effect in the cable will be the same along the entire length and unless the cable runs through areas of differing thermal insulation the same features of overheating will be apparent along its length. Where overloading is suggested as a fire cause there may be some section of the cable, which has survived the fire with minimal damage to its insulation. This should show the effects of the extreme overload (see later). Usually, a suspected cable submitted to the laboratory only shows the effects of overheating in the damaged insulation at the edges of the burnt area and shows no such effect where the insulation is undamaged by the fire. Under such circumstances, to say that there is *no evidence of an overload* would perhaps be an understatement of the position. The true position would be that *there is evidence that the cable was not overloaded*.

Resistive heating

This is the term normally used to describe the effect of a poor connection within an electrical connector. This may arise due to the screw terminal being loose or to the copper wire used being tarnished. A defective connection is produced which offers a slightly higher resistance to the normal circuit current flow. Overcoming this resistance causes heat to be generated in the connection during normal current flows, for which the circuit was designed and fused. Over a period, the high temperatures produced cause further oxidation or degradation of the connection. In some instances an apparently stable *glowing connection* is formed. Many workers have studied this phenomenon and temperatures in the region of 1250°C have been measured [7]. Alternatively, the connection may progressively overheat and burn itself out, or it may pyrolyse adjacent materials and start a fire. In the later stages the contacts may partly melt, or separate by a short distance allowing an electrical arc to form across the gap (*in-line arcing*, qv). As the circuit is carrying its normal current and no leakage out of the circuit is occurring, none of the normal circuit protective devices can detect the fault. If this type of fault occurs in certain types of insulators the heating effect could cause *tracking* (qv) between the phase and neutral connections, which may result in a fire.

Mains supply ('cut-out') fuses such as that in Figure 3.13 occasionally succumb to resistive heating and often show melted metal typical of in-line arcing.

In-line arcing

In-line arcing is the effect that occurs if a conductor carrying a current actually breaks and separates by a small distance. Providing that the separation distance is sufficiently small the

Figure 3.13 Supply company's main fuse ('cut-out') from a domestic installation showing the effects of overheating caused by poor contact between one terminal of the fuse carrier and the fuse receptacle (Photo: Twibell)

current flow is maintained by the generation of an electrical arc across the gap. Electrical arcs are exceedingly hot (of the order of 3500°C) and can readily melt and pyrolyse adjacent susceptible insulating materials. The pyrolysis products from rubber or PVC insulations can be easily ignited from the arc, which may also cause smouldering ignition of nearby cellulosic fabrics. As in the case of resistive heating, the circuit is carrying its normal current and none of the circuit protection systems currently in place can detect such a fault.

An example of the effects of in-line arcing is frequently seen in the UDF (unkinkable domestic flex) used with electric irons. During use of the iron the flex is being almost constantly flexed backwards and forwards and most of the stress is directed to the point where the flex enters the rubber sheath of the flex grip on the iron, or sometimes at the plug end. Eventually, the strands within a conductor will break. When the final one separates in-line arcing occurs. This frequently results in the generation of a small flaming fire at that point, which usually self-extinguishes when the supply is switched off. Householders are frequently puzzled that a fuse did not blow but an excessive current is not flowing and it is not until the fire burns through the insulation of an adjacent conductor that a short circuit can occur.

Airing cupboard fires

A number of fires occur in airing cupboards, often associated with the wiring to immersion heaters. Immersion heaters are usually rated at 3 kW, which means that they draw about 13 A in operation. A heat resistant flex normally carries the supply from the switched outlet to the heater in the tank. The householder usually stores clothing, etc. in the cupboard often to a high level and the flex is frequently disturbed, which might loosen the flex grip and damage the connections over a period. The combination of frequent disturbance, high currents, a high level of thermal insulation and the presence of flammable materials renders these locations particularly vulnerable to fault fires. On the other hand, such locations are similarly vulnerable to carelessness with smoking materials, and are an ideal location to start a slow developing deliberate fire.

Electric blanket fires

The most commonly encountered example of in-line arcing in terms of house fires is probably that of bed fires caused by electric blankets. Such faults presumably occur due to frequent flexing of the fine heating element within its plastic sheath and its eventual separation and the development of an in-line arc. The author has investigated many electric blanket fires. They are insidious in that the bed occupant may switch the blanket off or unplug it and get into bed unaware that a smouldering fire has already been generated. Figure 3.14 shows a photograph of the blanket lead from the scene of a fatal fire where an elderly woman died in her bungalow. The only fire damage in the house was to the bed and to the woman and her bed clothing. She had managed to crawl into her kitchen leaving a trail of small pieces of burnt nightdress and had collapsed and died in a corner. She was not a smoker. The photograph shows the smoke shadow of the blanket flex and plug demonstrating that the lead was unplugged during the period when the smoke was settling out. The electric blanket is the most obvious exception to the general rule that if a circuit or appliance can be shown to have been off or disconnected through the period of a fire, then it can be excluded from having been the cause.

Figure 3.14 Electrical blanket bed fire. Blanket mains lead moved to reveal smoke shadow showing that lead was unplugged during later stages of fire (Photo: Twibell)

Tracking

Some types of otherwise efficient insulating materials are susceptible to breakdown by high voltages under certain conditions. Thermosetting plastics such as bakelite are particularly vulnerable. Where such materials are used to separate terminals at mains voltage or higher, tracking may occur. This is often initiated by damp conditions, where a thin film of water condensing on the material will allow a small leakage current to flow between the terminals. Whilst the heating effect of this current tends to dry out the water and stop the current flow, it also appears to cause breakdown of the surface of the plastic, pyrolysing the material to leave a thin partly conductive track of carbon. Current flowing through this thin layer of carbon causes greater degradation of the plastic producing a better carbon track and a greater current flow. Whilst tracking can take weeks, months or years to actually commence, once a good pathway is established the process may rapidly degenerate until a high current flow develops through the damaged material. Such flows are often accompanied by electrical arcing, which may develop within part or most of the fault path. Initially the fault current will be limited by the resistance of the charred material but the flow may increase as arcing develops. Ultimately, it may burn out the fault path and extinguish the fault current or it may increase to a level at which a fuse in the circuit will blow. In any of these situations the arc could cause flaming ignition of the resultant pyrolysis gases or flaming or smouldering ignition of adjacent vulnerable materials.

Tracking may also be caused by resistance heating or in-line arcing within a defective connection. If the overheating connection is separated by the insulation from another conductor at a differing potential, the heat generated may eventually form a conductive track between the two conductors.

Faults in electricity intakes

Tracking occasionally occurs within the supply company's equipment, either within the plastic base of the meter where the incoming and outgoing leads ('tails') are connected, or within the combined Neutral and Live 'cut-out' fuse assembly. Often the originating fault can be seen to have been an overheating poor contact between the fuse contact pins and the receiver socket (Figure 3.13). This can result in a total burn-out of the incoming electrical installation (and loss of supply) and may cause the ignition of adjacent materials. In a three-phase supply, a defective contact in one fuse holder may result in tracking to one or both of the adjacent phase fuse holders. This may result in all three phases being involved in the ensuing burn out, with a consequent loss of supply. In some cases it may result in one (or two) phase(s) remaining connected which could pose hazards to the unwary occupant or investigator.

Any fire apparently starting in the electricity intake and metering installation should be examined closely. If a bypass connection has been made to the meter to obtain an illicit supply it may have used sharpened probes inserted into the meter or cut-out fuse connections. Similarly 'Black box' devices were popularly used in the 1980s to wind meters backwards and these were attached via heavy duty probes into the phase connectors either side of the meter. Repeated use of such probes could loosen the connections rendering them more vulnerable to resistive heating. Under other circumstances the bypass or black box could also overheat and cause a fire.

High resistance faults and stray currents

Most materials used in building construction offer a very high resistance to the flow of a stray electrical current, but their resistance will fall considerably if they become damp. Wood is a reasonable insulator when dry but can readily conduct when wet. Figure 3.15 shows a slide from an experiment where a 230 V mains supply was connected across a piece of wet wood. A current of about 0.4 A is flowing through the wet surface layers of the wood. The most noticeable effect was the generation of small electrical arcs within the wood surface during the experiment. These can be seen as small flashes in the still photograph. The current tended to boil the water out of the wood within a minute or so and further water was added as required. As the experiment proceeded more of the wood became charred in the tracking process and occasional spurts of flame emanated from the deeper burns.

A number of fires have been caused by conduction through wet wood. One such case in the 1980s was such a good example that it was used in demonstrations thereafter. It involved an earthed copper water pipe and a mains cable running parallel, a short distance away along a chipboard partition. A cable clip used to secure the cable had been nailed such that the nail passed into the cable and contacted the Live conductor. This situation is quite stable whilst the woodchip is dry. At some stage shortly before the fire a leak occurred in the pipe and the chipboard became wet. Once the water soaked into the chipboard between the water pipe and the Live nail, conduction through the wet chipboard caused ignition of the material. It should be noted that in the laboratory demonstrations water had to be added at regular intervals to sustain the current flow and the breakdown of the wood until the damage was sufficient to be self-sustaining. It was the continuous leakage of water from the pipe in the case situation, which promoted the fire.

Figure 3.15 Tracking through wet wood at 230 V AC. Mini arcs can be seen within the wet wood. Dark areas show where carbonisation of the wood has already occurred (Photo: APU/LFB)

Old timber beams that have been attacked by fungi usually show high conductivity. Other building materials can be similarly conductive under certain conditions but the fault may not cause a fire straight away. Many building fabrics tend to absorb water from damp atmospheres or give them up to dry atmospheres, and hence their conductivity will vary with weather conditions or on the rise or fall of the water table. Hence this type of fault may occur in the summer but not cause a fire until climatic changes occur. Such a fault requires that an uninsulated Live conductor comes into contact with the fabric of the building, which should not happen in normal circumstances. If there was an RCD in the circuit this should operate to disconnect the fault path before any appreciable current flows.

Neutral–Earth faults

Although the Neutral of the substation transformer is connected to earth at the substation, the Neutral of a particular householders supply may be a few volts above earth where it enters the premises. A simplistic reasoning for this can be seen in the following example and by reference to Figure 3.16. The supply cables distributing the supply to a number of consumers have a finite resistance, which gives rise to a voltage drop across each section of the conductors. Suppose that there is a total consumption on the circuit of about 10 kW (say 40 A) which is being drawn at or beyond Consumer 3 in Figure 3.16. The 40 A being drawn through the Neutral cable from the substation to Consumer 1 would give a voltage drop in this section of cable of ($V = IR$) $40 \times 0.2 = 8$ V. Thus the Neutral supply at Consumer 1 would be approximately 8 V above the Earth potential at that point. Similarly, for Consumer 3 the Neutral will effectively be 16 V above the Earth. There will thus be significant measurable voltages between the Earth and Neutral conductors within their premises. These will be added to by contributions from other consumers on other phases sharing the same Neutral line. To avoid excessive Neutral–Earth potentials being generated, the supply company may connect the Neutral to Earth again outside each consumer's premises. Such installations are sometimes called PME (Protective Multiple Earth).

Neutral–Earth potentials of a few volts may enable a current of several amps to be drawn between the two conductors, and flash-lamp bulbs can often be illuminated if connected between them. Clearly, if the two conductors are not fully insulated from each other and some piece of low resistance material such as a metal wire off-cut, metal swarf or similar

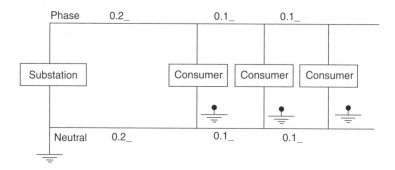

Figure 3.16   Generation of Neutral–Earth potentials (see text)

material could bridge the gap, a fire could result, but only if suitable flammable material was also present at that location.

Electrical sparks and arcs

The terms 'spark' and 'arc' are often used interchangeably although there are subtle differences between them. The term 'spark' is used to describe both a transient electrical discharge through air or a gas, and a high temperature incandescent particle. High temperature incandescent particles can be formed by friction effects (e.g. the impact of flint and steel) or may be ejected from a high temperature electrical event such as a short-circuit. An electrical spark is a short-term phenomenon in which a discharge occurs through an air gap between two items of differing electrical potential. The difference in potential has to be sufficiently large to break down the insulation resistance of the air, which for dry air is of the order of 30,000 V/cm. When the breakdown voltage is exceeded the air in the gap ionises and a flow of current occurs as a visible high temperature discharge. Sparks emit a range of electromagnetic radiation including light, heat and radio frequency interference and the explosive heating effect causes a noticeable sound. Sparks usually last for a very short period (fractions of a second) sufficient to cause near equalisation in potential between the two charged items. Electrostatic sparks built up by friction between dissimilar substances tend to be of short duration and are often of low energy, depending upon the circumstances and the scale of the event. The obvious extreme case is lightning where massive charges are built up due to air movements in clouds and many millions of volts generate sparks of lightning perhaps thousands of metres in length and currents of thousands of amps can flow.

Electrical arcs are essentially continuous sparks where the conduction across the gap is maintained for much longer periods than with the transient nature of a spark. If two contacts within a circuit carrying current are opened slowly, at the moment of separation the voltage developed across the very small gap is likely to be sufficient to ionise the air in the gap and conduction occurs across the gap in the form of a spark. Providing that the separation distance remains small, the continuing potential will maintain the spark as a continuous arc. In AC circuits the voltage across the gap reverses through zero a hundred times per second (50 Hz) effectively stopping the arc but within the timescale, and if the distance is short, the arc re-establishes each time that the voltage across the gap builds up. The result is an apparently continuous arc which continues until either the separation distance increases beyond the breakdown limit or the supply is discontinued.

Sparks and arcs are very high temperature phenomena and operate at well over 1000°C, but the transient nature of a spark means that it may not release much energy overall. There is a minimum energy required to ignite flammable gas or vapour mixtures in air, and some sparks may not have the requisite energy content. Arcs clearly release much more energy, produce high temperatures at the arcing points and will clearly ignite many gaseous mixtures, or even finely divided dust mixtures in air. Depending upon the circumstances an arc may well pyrolyse adjacent materials and ignite the gaseous products.

Arcing can develop in wiring where the insulation is damaged but at ordinary temperatures requires some further influence such as ingress of moisture. If the moisture is boiled out the arcing should cease. If the insulation is being damaged by severe overloading or by heating in the fire, then arcing readily occurs at elevated temperatures [3]. It appears that the products of combustion may be more readily ionised than dry air. (See also 'Tracking in burning wood/tracking through char'.)

Ignition of gases and vapours by electrical arcs

Electrical arcs and many sparks have sufficient energy to ignite flammable gases and vapours providing that they are mixed in the correct proportions in air at the spark gap location. A fuel may already be present due to an escape of gas or to the spreading of a highly volatile accelerant such as petrol. There have been a number of recent fire murders where petrol has been spread in premises or over a person and has then been ignited by the perpetrators. Often when caught they will accept the spreading of the petrol but not its actual ignition and they may then claim that the ignition must have been accidental due to static electricity or some other stray electrical spark. The author has investigated a number of such situations and much of the investigation has been directed towards proving that there was no accidental means of igniting the petrol vapour within the scene. If a person is doused in petrol in a building there is the possibility that they might ignite themselves if they have acquired a static charge by, for example, walking on a synthetic carpet and then discharging themselves by touching an earthed metallic item. If there is no such item, which could conduct the charge away, in the vicinity where ignition occurred, then the possibility would appear to me to be remote. An alternative might be that they reached out and switched a light switch on or off. This again appears to me to be an unlikely scenario as petrol vapour mixed with air in the correct proportions must enter the switch. Although ordinary light switches are not intrinsically safe, it might take a while for diffusion from the petrol soaked hand to achieve the correct mixture within the switch contacts.

With a large volume of petrol spread within a building there is the possibility that the vapour might be ignited by someone inadvertently switching a switch, or alternatively by some automatic electrical switching such as the operation of a thermostat or of the starter mechanism of a fridge or freezer. Petrol vapour is much denser than air and mixes relatively slowly with it and the mixture builds up from low level. If ignition was from a light switch, there would have to have been such a large volume of flammable mixture built up that a major explosion would result. In several cases investigated by the author the petrol had been spread in areas of the building away from the kitchen or utility room where items such as fridges and freezers were located. Again if the petrol vapour mixture was sufficient to produce the correct level for ignition at the motor contacts, a large volume of flammable mixture would be located elsewhere, with the consequent likelihood of a low level explosion.

Most fridges and freezers have a contactor mounted on the side of the metal housing of the motor/compressor unit. Whenever the fridge starts the contactor rapidly cuts in and out to feed a short duration supply to a secondary winding on the motor, to give a sufficient kick to the motor to start it against the resistance of the compressor. Operation of the contactor will almost certainly produce electrical sparks. If a flammable gas or vapour (of appropriate mixture strength) is present in the region of the contactor then ignition may occur. One fortuitous feature of fridges and freezers is that the location of the motor compressor units and contactors are often relatively greasy and attract dust and large hairs and fibres (Figure 3.17). These can be examined to determine whether there are any signs of singeing which might have been caused by the ignition of flammable gases.

Loss of Neutral link

Occasionally, the Neutral return to the substation may break or burn out. In some instances it may perhaps not be reconnected after maintenance work on the system. In a three-phase

Figure 3.17 Typical accumulation of fluff around starter contactor of fridge compressor (Photo: Twibell)

distribution system this does not result in a complete loss of supply, as the three phases are interconnected via any loads that the various consumers may have connected. Figure 3.18 shows this situation. With the link to the centre point of the substation transformer broken, there is nothing to stabilise the Neutral mid-way between the three phases. If it is no longer pegged to earth, it will effectively drift around at voltages above Earth somewhere between those of the three phases dependent upon the various loads being switched in and out by the consumers on differing phases. If it remains pegged to Earth, then the three phase potentials will effectively drift relative to the earthed Neutral by the same effect. Either way there is no effective stabilisation of the supply within the normal 230 V tolerances. If there is a large load on one phase, coupled through the neutral line to a very small load on another phase there will be an uncorrectable imbalance in the distribution voltages. A consumer on the phase which has a very large load connected will see a very low voltage across his installation whilst the one on the low load phase will see a very high supply voltage. This high voltage is likely to overrun and burn out incandescent lamps and may cause small appliances to overheat and generate fires. The author has been informed of such a fault occurring in a factory situation, which resulted in many of the emergency lighting units in the premises catching fire.

Misuse of electrical appliances

Many fires are caused by misuse of electrical equipment rather than by true electrical faults and are beyond the scope of this chapter. Clearly, if smoulder susceptible materials are placed

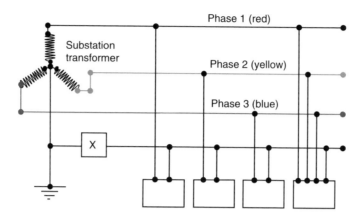

Figure 3.18 Local supply with disconnected Neutral return. Current flows from phase to phase via the loads of the various consumers and the interconnecting Neutrals. Without the Neutral link back to the substation transformer the supply voltages at each consumers cannot be stabilised (see text)

too close to any hot item, electrical or otherwise a fire could result. The Windsor Castle fire was caused by a hot halogen lamp close to a curtain. Whilst they may not heat to such a high temperature as halogen lamps, ordinary incandescent bulbs can initiate smouldering combustion in susceptible materials. In one case investigated by the author, a fire started in a stack of rubber exercise mats in a Young offender's institution. Arson was suspected but the fire had started at a high level in the stack of mats in the large walk-in storage cupboard. The mats had been stacked almost up to ceiling height and had pushed the incandescent light bulb fitting upwards towards the ceiling. Smoke patterns showed that the switch had been on during the fire, but the bulb had broken during the fire recovery operations. Tests on the rubber of the mats with various sized light bulbs, showed that bulbs of 60 W and over could initiate a smouldering fire in the foam rubber if in contact with the mats.

It is obvious that a chip pan will catch fire on a gas ring if the oil boils and the heavy vapour or the oil can spill over the edge of the pan into the flames. It may be less obvious that ignition can also occur on a non-glowing electric ring. The autoignition temperatures of cooking oils are of the order of 450°C. In tests carried out by the author on a non-glowing metal element, flaming ignition of rape oil occurred at oil temperatures below 420°C in some instances, but ignition did not always occur. Thus ignition is not absolutely certain and whether or not it occurred would depend upon the pertaining circumstances. However, the results of many a fatal fire investigations involving drunken persons and chip pans adequately testify to the frequency with which this form of ignition occurs.

Many fires are caused by misuse of electrical appliances, particularly heaters. The drying of clothing with electric heaters has been a particular problem over the years. Covering a convector heater with cellulosic textiles or paper can give rise to a fire as the self-ignition temperatures of these substances are very low. Unless the heater is a modern one, which has a thermal cut-out to prevent the element overheating, temperatures of this order could be achieved in a covered heater within a short period.

Clearly, a radiant bar electric heater is a dangerous item if clothing is placed too near to it and cellulosic materials which fall onto the front guard are likely to be ignited. If such a heater is knocked over so that it falls face down on a carpeted or wooden floor, a fire can be generated in a number of minutes.

False attributions

A number of situations or observations are frequently misinterpreted by untrained investigators. As discussed earlier the most common is that of the cable, which ran through or close to the fire seat area and which is partly burnt or may be showing 'arc' damage, 'short circuit' damage or 'electrical activity'. Over a number of years I have examined numerous examples of such pieces of cable submitted to the laboratory, and most, if not all, can be eliminated as a fire cause. The effects found are usually typical of a fire having burnt through the cable and any severe arcing or short circuit damage merely shows that the cable was energised when this happened. The signs of electrical activity in the cable may be due to a short circuit, or may be due to arcing/tracking through char.

Tracking in burning wood/tracking through char

A similar effect to tracking often occurs during a fire, and is a result of the fire rather than a cause of it. If the conductors are still energised after their insulation has burnt away they may come into contact with charring woodwork. A sudden arc can occur through the char resulting in a burst of flame from the area. Photographs from a demonstration of this phenomenon are shown in Figure 3.19. Such an arc may last for a fraction of a second or for a few seconds before extinguishing. The separation distance between the conductors is usually far greater than that which would support a stable arc in air and the low resistance of the char is clearly supporting the arc and forming part of the circuit. Depending upon the current flow and or its duration the effect may cause partial melting of the two conductors or may cause them to melt and sever. If they sever large melt globules often form on the exposed ends. Unless the conductors actually touch each other, the effect is not quite the same as a short circuit as the current flow in the arc is limited to some extent by the resistance of the char. The current flow may thus not be sufficient to rapidly blow the circuit fuse and several such char arcs may occur at the same or different points in the circuit before the

Figure 3.19 Arcing through char. Woodblock heated by blowlamp and 230 V cables applied to hot charring surface results in sudden massive arcing between contacts (Photo APU/LFB)

supply becomes disconnected. Due to the power dissipated during these arcing effects and their duration, the amount of copper melted during this phenomenon may be much greater than that formed in a typical full short circuit. The resultant appearance of the melted conductors at the scene may be misleading to the untrained observer.

Damaged conductors and aged insulation

It has sometimes been suggested that the fire was due to the cable conductor having been damaged at some point, perhaps partly cut through with a knife or that the conductor had been stretched in pulling the cable through a conduit or round a tight bend. Béland in Canada carried out a number of tests of such circumstances. He found that the localised heating effect at a small damage point was easily contained by the thermal mass of the adjacent sections of conductor, although the effect was greater if a number of strands of a stranded conductor were cut. He also found that copper conductors could be stretched such that the cross section of the conductor reduced by 20% before the conductor broke, but that this resulted in a temperature rise of about 1.2 times the temperature rise of the unstretched cable. He finally concluded that 'dangerous overheating of a conductor is possible only under the most extreme conditions unlikely to be met in practice. The damaged conductor as a cause of fire is a gross exaggeration' [8]. Similar findings were made by Ettling [21].

Old wiring is occasionally alluded to as the cause of the fire, but this may be a contributory factor rather than a direct cause. Even if the insulation has become brittle and is flaking away, the conductors should still be sufficiently well separated for insulation to be maintained unless some other factor operates. Thus, if some force acts upon the conductors to press them together (which would most likely cause a short circuit) or the cable becomes damp, then a fault may develop. In his experiments Béland [9] found that old cables with brittle insulation were no more prone to arcing at normal temperatures than was modern wiring. In the absence of other evidence old wiring cannot be blamed as the direct cause of a fire.

Overfusing

On occasions it has been suggested to me that the fire was caused by the householder having put too large a rating of fuse in the fuseholder, or in one case a hairgrip and another a nail! In none of these cases can this be deemed to have been the direct cause of the fire: some other fault would have to have occurred in the overfused circuit to cause a fire. An inspection of cables from the suspect circuit usually shows that it has not been overloaded.

Electrical examinations at the scene

Ensuring electrical safety

A particular emphasis must be made with respect to electrical dangers at fire scenes. It is essential that the investigators can work in the secure knowledge that the supply has been properly disconnected and cannot be reconnected by other persons during the examination. The Electricity at Work Regulations 1989 apply to any environment where electricity is used or available and where electrical hazards may occur, and certainly apply at scenes where the supply is not isolated. Regulation 3 makes it the duty of any employer or employee or self-employed person to comply with the provisions.

In house fires where much of the building remains standing, the Fire Service have usually disconnected the supply by switching off the main isolator. (Often they have switched all MCBs off in the consumer unit as well which is to the detriment of the examination.) If the electricity supply company's service cable entry has been damaged the company may have already been out and disconnected the supply in the street. It is imperative that the investigator checks the position himself before starting the investigation and takes appropriate action. If the service entry is undamaged and the supply still connected I would remove the fuse or fuses (labelling and retaining them for subsequent possible examination) and replace the empty holders in the fuse boxes to cover the otherwise exposed terminals.

Regulation 13 relates to 'Precautions for work on equipment made dead' and requires that adequate precautions shall be taken to prevent the equipment being re-energised during the work. If the service entry is damaged and Live, the supply company should be contacted to make it safe. It is possible to work around such Live service entries but the dangers inherent in working alongside a supply which may have only perhaps the 300 A substation fuse between you and eternity should be evident.

In this respect, Regulation 14 relates to working near Live conductors but does give an opt-out if it is unreasonable in the circumstances for the supply to be dead. Regulation 29 gives a defence against contravention if the person can prove that he took all reasonable steps and exercised due diligence to avoid commission of the offence. However, Regulation 15 relates to adequate working space, means of access and lighting provisions being made around electrical hazards, which may pose problems at a scene, and Regulation 16 relates to the technical knowledge or experience or degree of supervision required under such circumstances. Thus, under the Electricity at Work Regulations 1989, if you are in charge of the investigation (or can be shown to be the one with most knowledge), you are requiring yourself and others to work in a dangerous electrical environment. If anything goes wrong you will be liable to prosecution.

Larger premises such as factories, schools, offices, etc. may present a different problem in that often only part of the premises is burnt and the works electrician or contractor may be brought in to make certain that the appropriate areas have been disconnected. Often this will be from an isolator, switchboard or fuse board in some other undamaged part of the premises. In factory situations there is often pressure to continue with business operations and the company continues to work. They therefore require to reinstate as soon as possible any electricity supplies in the unaffected areas, which may have been cut by the fire damage. There is a distinct danger that some other electrician or workman who is unaware of the investigation going on elsewhere might try to reconnect an affected circuit and inadvertently power up the damaged area under investigation. It is therefore imperative that when the appropriate electrical circuits have been isolated, no one can enter the switchboard and reconnect them. If this possibility exists it is essential that adequate safety warnings are displayed on the appropriate box or boxes, together if possible with some means of temporarily locking the cabinets.

Test equipment and its use

In addition to the routine equipment carried by the investigator it will be necessary to carry the following.

- A Voltstick or similar pen style (non-contact) probe, which indicates the presence of Live conductors through intact insulation. This must be tested on the day of the visit and the batteries replaced regularly.

- A well-insulated AC/DC autoranging neon tester or testmeter. Such a device (when connected between a conductor and a secure earth or neutral) will confirm the presence of any dangerous voltage. This overcomes such problems as the operator inadvertently testing AC mains on a DC range and obtaining a zero reading.
- A digital multimeter with mains and higher voltage ranges (AC and DC) and several low resistance ranges.
- An audible continuity checker (or facility on the multimeter).
- An Earth loop impedance tester, and an RCD tester may be required at some scenes.
- A selection of suitably insulated screwdrivers, insulated pliers and wire cutters, together with other tools.
- A triangular key of the type used to open external electricity meter boxes.

It is essential that the investigator is familiar with the test equipment and is aware of its limitations and any potential to give erroneous readings. Modern digital voltmeters tend to have very high input impedance and might indicate the apparent presence of very high voltages in some circumstances. If this is a stray-induced voltage it is likely that the current available is so low (microamps) that someone completing the circuit would probably not feel anything and the reading would drop to an apparent zero if measured on a low impedance meter. The investigator must be capable of distinguishing between such harmless stray voltages and potentially fatal high voltages, which may have resulted from a fault caused during the fire.

Continuity testers give an audible response over a low resistance range (perhaps up to about 80 Ω or so) and not just for direct contact. Many of these will indicate continuity through light bulb filaments or other low resistance electrical equipment. This can give misleading indications if the tester is applied to switch contacts if at least one terminal is not disconnected from the circuit (see 'Light Switches'). Similarly fuses must be disconnected from the circuit before testing. Even a fuse in a burnt 13 A plug top at a scene cannot be properly tested without either removing the fuse from the plug or at least removing the plug from the socket to break the continuity through other circuits connected, or shorts caused by the fire. The implications to other evidential materials before disconnecting or removing items from the circuit must also be considered.

Testing of cables or appliances

If any electrical testing is to be carried out at the scene (or in the laboratory) use only low voltage test apparatus for the initial tests. Insulation Resistance Testers (also known as 'Meggers') or similar high voltage test equipment should not be used indiscriminately to test for insulation breakdown. These instruments apply a high voltage (usually about 500 V) to the insulation under test. It can almost be guaranteed that if the insulation was severely damaged by the fire *after* the supply was disconnected, then the use of one of these machines will finally break down the insulation and produce a low reading. They can be useful as part of a careful examination procedure. In a number of fires it has been possible to show that the burnt wiring produced adequate insulation at low voltages but that on applying a high voltage from a Megger, breakdown occurred such that subsequent testing with a low voltage resistance meter gave a much lower resistance. If this occurs it must be clear that the final fire damage to the insulation was achieved *after* the supply had been disconnected.

Scene examination

The investigator will use his or her knowledge of fire investigation to deduce where the seat or seats of fire appear to be and to locate any electrical cable, appliance or other item to determine whether or not a fault could have been the cause. In general, if it can be shown that the cable or appliance was not energised at the time of the fire, then it could not have been the cause. (As discussed earlier the only likely exception to this rule is the electric blanket.) It is important to examine such items carefully even when the cause is obviously due to deliberate ignition. If an accelerant such as petrol is spread around the scene it is also important to establish whether or not there are any sources of accidental ignition.

An examination of the seat of fire region or (where the damage is too great) the room or area of likely origin, will be conducted. The investigator will clearly be undertaking a thorough investigation of all aspects of the scene rather than concentrating on just the electrics. Electrical items, cables flexes and appliances in the seat area will need to be carefully examined for elimination or to be identified for subsequent recovery and laboratory examination. Any lighting or ring main cable above the seat area will need to be examined for signs of electrical activity and recovered if necessary. A cursory examination of such wiring might show that the features are typical of a hot fire at that location having melted the conductors or caused arcing through char during the fire rather than having been the cause. It is essential that the possibility that this might have started the fire be considered and dismissed or put into context in one's notes.

The appropriate fuses in any final circuit under investigation should be removed and tested for continuity at the scene and, if required, recovered for further examination (see *Test equipment and its use*). Other fuses in adjacent appliance plugs may be relevant in showing whether or not the circuit was energised at the time of the fire. Fuses further back in the same circuit should also be tested and recovered as necessary. It will be clear from the scene examination whether or not any blown plug-top fuses involved will have been subjected to severe heating which will assist in the subsequent interpretation. An obvious drawback from any such investigation is that the fuse (of whatever type) may have been blown (or MCB tripped) prior to the fire. The investigator will have to seek information from witnesses as to the state of things prior to the fire, and this may not be reliable. In the case of rewirable fuses they can be examined and observed to have either blown, or to still be intact and there is little else which can be done with them. Spattered copper deposits on the carrier or fuse holder may be from previous events.

In some situations it may be suggested by the initial investigators that the fire originated within electrical switchgear, perhaps due to a sticking contactor or due to tracking across some fitting. A brief examination of the switchgear should be made at the scene and the item recovered for a later laboratory examination if necessary. It may be possible to negate such suggestions at a later stage.

Electrical items

Modern televisions and radios are much safer and less prone to catching fire than earlier versions, but the number of these and associated or similar items (e.g. video recorders, satellite decoders and computer equipment) is proliferating in homes and offices. Many of these items are left unattended but energised at the flex or often in standby modes. An examination of the plug fuse might show that it is still intact consistent with the fire having burnt through the electrics elsewhere and disconnected the supply before the fire reached the appliance or its flex. Fire damage to the item may be complete such that all combustible materials or components have been destroyed. In this case the only evidence discernible from the remains may be signs of electrical activity that would indicate that the item was energised at the time of the fire. The lack of such evidence would not necessarily prove the opposite. On the other hand, fire damage might be so minor that the item can be reconnected and shown to function normally.

The majority of items examined lie between these extremes, being too badly damaged to function but not completely burnt out. Electronic goods such as radios, TVs, video recorders and stereo systems have plastic cases, which melt and eventually burn in fires. The orientation of the item during the fire can usually be read from the direction of dripping plastics and waxes. If the item fell over during the fire this should show in a change in drip direction.

A fire external to the item initially causes a directional heating effect, which may be lost later as the fire spreads. The plastic case usually collapses down onto the internal components from the fire direction. If the item itself or the plastic case does not ignite, the results are usually characteristic. In a fire of short duration the interior electronics are likely to be sealed and preserved by the melted plastic. If, however, the item catches fire due to an internal fault the fire is likely to burn a hole in the case and the surrounding plastic then collapses in and around the hole. The resultant fire is likely to spread to consume more of the casing and to other items in the room. If an internal fire consumes the item the appearance of the remnants may be similar to that of an external fire burning the item and the examiner will have to rely on other features such as local burning patterns and on findings from the mains leads, plugs, sockets or fuses. Fires have often been blamed on television sets or stereo equipment, which have later been shown to have been switched off or not plugged in at the wall socket. The state of the internal mains switch can provide useful information. Instrument fuses within a piece of electronic equipment may not provide reliable evidence. They are similar to plug top fuses in construction but use glass envelopes and have no sand filling. External heat readily penetrates the glass making them more susceptible to thermal rupture. This is particularly so with the partly spirally wound delay type of fuse.

Many pieces of electrical equipment such as washing machines, fridges and fluorescent lighting units are housed in metal cases. The metal case often provides more protection from an external fire source than does a plastic case, and can still show directional effects due to damage to the external paintwork or by smoke staining. Unless the external fire has been very intense or of long duration, the internal components may be relatively undamaged. Internal wiring adjacent to the hot casing will clearly show damage and may eventually catch fire. In these circumstances, it may again be difficult to determine whether the fire damage was internal or external. Wound components such as motors, chokes and transformers contain a relatively large mass of iron laminations, which have a high thermal inertia. External fire damage to such items usually commences as localised damage to exposed windings. The mass of iron initially acts as a heat sink preventing other parts of the winding from heating. This contrasts with the effect of internal heating caused by a fault in the windings where heat is generated uniformly within the faulty winding resulting in the whole component cooking to destruction. Unfortunately, prolonged external heating will eventually penetrate deeper into the windings and heat the iron core resulting in a similar uniform cooking of the component. (Electrical tests can be carried out on less well-burnt items but such testing is beyond the scope of this chapter.) Motors contain armature bearings that can dry out and seize. Clearly, free rotation should be checked but the heat of the fire can dry out the bearings and many bearing housings are cast from soft alloys which melt.

Signs of electrical activity in burnt conductors

Cable conductors in relevant parts of circuits will need to be examined for signs of electrical activity during the fire. It can be very difficult to fully interpret damage to conductors and the following notes can only be indicative. The appearance of any melted or arced conductors can often be very different dependent upon the circumstances which caused the effect, and may be further confused by subsequent heating during the fire.

In most cases, the conductors will have shorted due to the fire having burnt through the energised cable and the results will be a relatively small portion of melting at the same point on two of the conductors. A third conductor in the same cable is likely to be unaffected. Here, the short circuit will have been of very short duration before the fuse blew. The appearance of the conductors after a short circuit may depend upon the whereabouts in the distribution circuit in which the short occurred. If it occurred within a final circuit, the energy dissipated and the effect obtained will be typically less than if it occurred in an earlier part of the circuit in more substantial wiring. In a final circuit the resistances of the cables or flexes and of the later circuit fuses will reduce the available current and the lower rated fuses may disconnect the circuit slightly earlier into the fault.

The typical short circuit result will be that the conductors are clean and show their original microscopic draw marks right up to the fault point. One conductor may show just a smooth gouge in its section, whereas the other one may show melted copper at a corresponding location. Multistranded conductors will show damage at only one or few of the strands, the others remaining undamaged and pristine. In some instances, melted copper will be found on both conductors at the same location. Often one conductor is completely severed by the event. In a number of short circuits created by the author in final circuits, one conductor became severed with the severed supply end remaining free whilst the other (load) end fused to the other conductor. This suggests that the short blew itself out at the same time that the fuse blew.

If arcing occurred within the cable the results would be similar except that more extensive melting of the conductors would be likely. This is because the fault current will have been limited by the arc (and may not have blown the circuit fuse), and would have continued over a longer period. Again it would be expected that the unaffected areas of the conductors would remain appreciably below the melting point and draw marks would still be present. If the condition developed into arcing through wood char more extensive melting of the conductors would occur. Again severing of the conductors is a likely outcome and the melting will be likely to extend through the conductor cross section. Large droplets of melted copper can be formed and the heat of the event is likely to affect, and bring near to melting, a short length of the conductor. The effect may not be so pronounced as with fire melting though and the draw marks on the conductors are likely to survive closer up to the damaged area. Multistranded conductors will show melting together but again there is likely to be a relatively sharp transition.

If the fire burnt through the cable and melted the unenergised conductors later in the fire, the appearance could be very different. Conductors which have been at temperatures close to their melting point for long periods often take on a more crystalline appearance and can become very brittle. In these cases, the draw marks on the conductors may be lost and blisters may form on the surface. Lengths of conductors reaching the melting point start to flow and large droplets of copper can form encompassing the entire conductor section and finally severing the wire. There are likely to be relatively long lengths of the cable, upwards from a few centimetres, which show the semi-melting effect. Where there are more than two conductors all conductors in the same environment should show the same effect. Multistranded conductors will show the effect on all the strands which will be seen to be merging into the melt. In many instances the reason for the fire having been particularly hot at the location may be deducible. Thus the cable may have been at a location in a ceiling where the fire first broke through into an upper floor level. It is possible that later fire effects such as these might be superimposed onto the effects of an earlier short circuit event, which could complicate the interpretation. During a fire, the effects of structural collapse often causes heated conductors to break through physical forces rather than from electrical activity. Béland [24] has shown that the appearances of the broken ends depends upon the temperature to which they were exposed at the time of the break.

Further problems in interpretation can arise where copper conductors exposed by the fire have come into contact with some other metal such as aluminium. If a lower melting point metal can melt onto the copper conductors and alloy with them, an effect similar to the 'M' effect in fuses occurs and the metal effectively eats through and melts the conductor at a lower temperature than that of pure copper. This produces beads or droplets on the conductors, which might be mistaken for electrical activity [19]. Again though the extent of the melting and appearance of the conductors approaching the melted portion is usually not what would be expected from electrical activity. Interactions between copper and aluminium in particular have been widely reported from fire scenes. These are a result not a cause of the fire.

It should be clear from all these that it can be very difficult to interpret the types of fault which occurred from the burnt wiring remnants and that the investigator will need considerable experience to interpret some cases. Further information can be found in references [2,3,9,10,24].

Did the fault cause the fire or the fire cause the fault?

The difficulties in interpreting damage to conductors has led various workers to try to find more scientific methods which might categorically answer the posed question. Thus it might be possible to show that the conductors were being heated by some other fault (rather than by the fire) prior to the short. Such methods involved examining the conductors at or in the vicinity of the melted features. Some have used Auger Electron Spectroscopy (Auger analysis) to analyse the copper to determine the levels of oxygen or other non-metallic elements picked up during the heating process [11,12], and others have used SEM to examine the microcrystalline structure [13,14]. None of these methods appears to have been widely adopted, and the analysis of arc beads by Auger Electron Spectroscopy appears to have been largely discredited [15–17]. (The controversy continued however at the 16th IAFS meeting at Montpellier in 2002, and perhaps it would be fairer to describe such techniques as *non-proven* at the time of publishing.)

Signs of overload in cables

When a cable is seriously overloaded the heat generated in the conductors raises their temperatures rapidly to beyond the capabilities of the insulation. The result in many insulation types is to start to pyrolyse the insulation and to generate bubbles of pyrolysis products or to vaporise the plasticiser. The result is that bubbles will start to form in the insulation closest to the conductors. In thermoplastic insulation in particular, the insulation will soften and the gases released may form a void between the insulation and the conductor. This effect can often be recognised after the event as the insulation becomes easier to strip from the conductors and is sometimes called '*sleeving*'. If the fault current persists the outer insulation also starts to pyrolyse and bubbles form in that also. The progression is shown in Figure 3.20. Dissection of the cable in the laboratory should show that the heating is uniform along a significant length of the cable. Cables from fires frequently show this type of effect at the fire damaged edges where the fire was destroying the insulation but the effect only penetrates a short distance into the undamaged insulation. Clearly copper conductors are good conductors of heat and the extent to which the damage is seen will depend upon the size of the conductors, but such effects are likely to penetrate for a few inches to a foot or so into the undamaged cable. Contrast this with a genuine overload where the same effect will be seen along the entire length of the cable that was in the same thermal environment.

Any cables removed from the scene will require adequate labelling and description. It is essential that the supply and load ends of the cable are clearly marked and all relevant information given. If the cable forms part of a ring main circuit the ends should still be identified as it may subsequently be shown that there was a fault which rendered the ring incomplete.

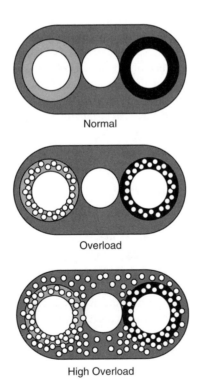

Normal

Overload

High Overload

Figure 3.20 Progression of overload in T&E cable. Formation of vapour/pyrolysate bubbles in softened insulation

In the laboratory the cable can be sectioned at intervals to show the extent of any internal heating effect or to show that the effect disappears away from the fire damage. If necessary a better section may be achieved if the cable is cast in the resin (as used for switches) and subsequently sectioned and polished.

Heated connections

The appearance of overheated connections can be characteristic after the fire. Although many connectors have a lower melting point than copper, they often survive the fire. Where a connector or plug-socket connection has made poor contact this may be evident, by contrast with other connectors from the same component. Often the heated connection will show extensive degradation, loss of metal, surface deformations, oxide formation and blistering which is not present in adjacent connections.

Using fault tracing to deduce the course of a fire (Arc Fault Analysis)

When a fire occurs in a building it will eventually start to burn through energised circuits and cause short circuits or arcing and produce arc melt damage. This can be useful in tracing the course of the fire through the building. If the fire burns through the cable insulation and causes a short this often severs one or more of the conductors at this point. If the short burns itself out but does not operate a fuse or breaker, the cable beyond the fault and remote from the supply is now dead, whilst that towards the supply is still energised. The fire may then burn through the cable again causing a further short or arcing closer to the supply. If the circuit fuse now blows, the whole cable is now dead, but some other cable may be similarly affected, and other shorts may occur. If the pattern of faults and fuse blows is studied it may be possible to work out the sequence of fire attack on the electrical installation, which can give information on the location of the fire and its spread. Even if each short blows the circuit fuse and isolates that circuit, the overall information supplied from the wiring and fuses can be useful. Delplace and Vos [10] originally published in this area. More recent studies have been made by Svare in USA and Carey in UK [25].

Laboratory examination of fuses

Cartridge fuses should be X-ray photographed. This enables the remnants of the fuse wire or ribbon to be viewed through the cartridge and the sand packing, and reveals the location at which the fuse blew (Figure 3.21). The conditions required to obtain decent X-ray photographs on a particular X-ray instrument will require experimental determination.

The result for a cartridge fuse is highly diagnostic of the type of fault that caused the overcurrent. In a study of small BS 1362 plug top (13 A series) Twibell and Christie [6] found that at low overcurrent factors of up to about four times, the fuse blew at the edge of the M-effect bead. This is as would be expected from the desired effect of the M-effect material. The rupturing appeared to have occurred in an explosive manner with globules of metal sputtered into the sand around the break point. Where fuses of similar ratings and construction were wired in series in the test circuit, only one fuse tended to blow. The others appeared to show no obvious ill effects.

At higher overcurrent factors the fuse blew either side of and away from the edge of the bead. This is presumably due to the heating of the fuse wire being so rapid that the thermal inertia of the bead delays the heating of the centre section and the fuse ruptures explosively

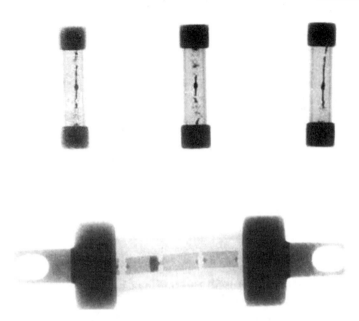

Figure 3.21 X-Ray photographs of plug-top fuses and larger cartridge fuse blown under very high current (short circuit) conditions

away from the bead. When a number of similar fuses were wired in series in the circuit the tendency again was for only one or sometimes two to blow. However, the surviving fuses (and the unblown length of fusewire in the blown fuses) showed deformations, which could be used as a diagnostic aid. These deformations appeared as adjacent areas of swelling and thinning of the fusewire, suggestive of their being about to blow at the time that the circuit opened due to the break elsewhere. This effect was described as '*short circuit survival*' or '*SCS*'.

As the overcurrent factor increases to full short circuit conditions it was found that the fuse wire tended to rupture in both central locations either side of the bead. When a number of similar fuses were wired in series in the circuit the tendency was for all the fuses to blow on both sides, although occasionally only one side blew. Any surviving portion was likely to show SCS.

Similar effects can be seen with larger cartridge fuses with ribbon type elements. The constriction with the M-effect material can be recognised due to the increased density of the image at that location. At low overcurrent factors the element ruptures at the constriction with the M-effect material. At higher overcurrent factors it can be seen to blow at one of the other constrictions and at very high overcurrent factors it is likely to blow at more than one or all of the non-M-effect constrictions.

Thermal rupturing of plug-top fuses can be distinguished from low overcurrent blowing at the M-effect bead by X-ray examination. In electrical blows there was almost always an explosive appearance to the sputter particles and in all cases some sputter material had been moved away from the original line of the fuse wire. By contrast, thermal blows showed a non-explosive break in which the wire had usually thinned out and drawn to a point facing the bead, with no movement of material away from the line of the wire. Breaks of similar appearance normally occurred on both sides of the bead.

Switches

It may be crucial to an investigation to show whether or not a socket outlet switch was switched on at the time of the fire. It may be that, by proving that a switch was off, a suspected appliance or circuit could be shown not to have been the cause of a fire. Although outwardly appearing to be well burnt, the internal mechanism of socket switches or light switches may survive the fire with little change. In any event it is essential that no one should try the switch or otherwise interfere with it prior to its examination by a knowledgeable investigator. Appropriate electrical tests can often establish that the switch was ON but sometimes movement may have occurred within the mechanism allowing closed contacts to open slightly and misleading results can be obtained. Also if the contacts are tested using a high resistance range, conduction by the carbonised plastic may be observed. Any switch or switched item should be removed as carefully as possible (preferably within its wall mounting) and submitted to the laboratory.

Light switches

Often it is important to determine whether the lights were on or off in a particular room or location at the time of the fire. This sort of examination can only be carried out if the scene is relatively undamaged. With short duration fires the wiring in the walls and ceiling usually survives. In the case of pendant light fittings, the heat of the fire is likely to destroy these at a relatively early stage, but the heat may or may not damage the wall switches lower down. If the light was on it is likely that the pendant cable will have shorted during the fire and blown the fuse. If the pendant cable is still present it might be possible to prove that it was energised from the presence of a small electrical arc mark on the conductors. If the pendant and fitting has been destroyed it may still be a relatively easy job to determine whether the lamp was on or off providing it was controlled from a single switching position where the switch is only superficially burnt. Although there is a convention for which way up a switch should be, there are a number of reasons why this should not be relied upon and the switch should be tested with a continuity tester or resistance meter. In order to do this it is necessary to disconnect one of the connections to the switch. This is because if some other lamp in the same lighting circuit is switched on (or the circuit is effectively shorted elsewhere by the destruction of insulation) the tested switch may appear to show continuity in both switched positions. Note the original position of the switch and any smoke or fire patterns which would confirm that it has not been interfered with since the fire.

The testing of lighting circuits that can be switched from two or more locations is much more complicated. Here it will almost certainly be necessary to establish the exact nature of the switching and the switches involved and to use jumper leads to test across the entire circuit. The actual testing procedure is beyond the scope of this Chapter and the reader should refer to diagrams of two-way switching and cross-over circuits.

Laboratory examination of switches

If possible the continuity across the switch contacts should be tested using a low resistance range on the multimeter. If full continuity is found, clearly the switch was ON. A high resistance may be due to continuity through char.

If the switch appears to be an open circuit further tests should be made, as the spring holding the contacts closed may have been damaged by the fire. Locate the position of the contacts by X-ray photography so that the position to make saw cuts can be determined. Some socket switches have double pole switches, both contacts of which should be examined. The switch can cast and sectioned as in the method of Twibell and Lomas [18], using a clear casting polyester or epoxy resin. When the resin has properly hardened the switch can be sectioned through the switch (or switches) and polished to examine the rocker and contacts. The position of the internal remains of the plastic rocker or toggle should indicate which position the switch was left in prior to the fire. The electrical contacts should either confirm this or show that movement has occurred as a result of fire damage. In several cases apparently open circuit switches have been shown to have been ON at the start of the fire.

Reporting of fires

The investigator will write a report or statement on his findings. This should describe the main findings and should include an interpretation section where the findings can be discussed and put into context. The various possibilities for fire cause should be considered and the relative strengths of opinion relating to these possibilities should be assessed from the findings.

Even in the case of a deliberate fire any alternative possibilities should be discussed and put into context. Thus, if for example, the scene investigation reveals that a deliberate fire was set in one room, but due to fire-fighting activities an electrical item cannot be accounted for, the statement should acknowledge this. Assuming that other electrical or accidental causes can be dismissed, the interpretation will perhaps be '*In my opinion the fire was most likely caused by the direct naked flame ignition of materials in the room*'. This will be tempered by '*As....I cannot entirely exclude the possibility that an electrical fault* (within the missing item) *caused the fire although I consider this to be unlikely/highly unlikely*'.

Conclusion

It should be evident from this chapter that faults do occur in electrical installations, but that they do not necessarily lead to fires. In a properly constructed circuit most faults would be detected and rendered safe by circuit protection devices before sufficient heating had been generated. Short circuits do not spontaneously occur in lengths of wiring and if they did they should rapidly take out a fuse or other device.

The types of faults that do cause fires are largely effects like resistive heating and in-line arcing, which cannot at present be detected by circuit protective devices. Whether any electrical fault can cause a fire also depends upon the presence of suitable readily flammable material or vapours at the fault location. There can be no doubt that electricity does cause some fires but the number is likely to be far fewer than the statistics suggest. Electricity has acquired an undeserved bad reputation.

It is essential that fires are investigated by properly trained persons who have proved their competences. The wrongful attribution of so many fires as of electrical origin continues to be a drain on society.

Acknowledgements

The editor and author gratefully acknowledge N.J. Carey (London Fire Brigade) for his contribution to photographs 3.12, 3.15 and 3.19.

References

1   J.D. Twibell (1995) Reporting of electrical fires hides true picture. *Fire Prevention* 282 (September): 13–15.
2   B.V. Ettling (1986) A guide to interpreting damage to electrical wires. *Fire and Arson Investigator* 37(2): 46–47.
3   B. Béland (1980) Examination of electrical conductors following a fire. *Fire Technology* 16(4): 252–258.
4   BSI/IEE (2001) *Requirements for Electrical Installations, IEE Wiring Regulations*, 16th edition (BS 7671: 2001) ISBN 0 85296 988 0, IEE P.O. Box 96 Stevenage UK SG1 2SD.
5   Guidance Note 4, *Protection Against Fire*, 4th edition. ISBN 0 85296 992 9, IEE P.O. Box 96 Stevenage UK SG1 2SD.
6   J.D. Twibell and C.C. Christie (1995) The forensic examination of fuses. *Science & Justice* 35(2): 141–149.
7   Y. Hagimoto, K. Kinoshita and T. Hagiwara (1988) Phenomenon of glow at the electrical contacts of copper wires. *NRIPS (Japan) Reports – Research on Forensic Science* 41(3), a translation abstract from an Australian web source TC Forensics (http://members.ozemail.com.au/~tcforen/japan/3.html).
8   B. Béland (1982) Heating of damaged conductors. *Fire Technology* 18(3): 229–236.
9   B. Béland (1982) Considerations on arcing as a fire cause. *Fire Technology* 18(2): 188–202.
10 M. Delplace and E. Vos (1983) Electric short circuits help the investigator determine where the fire started. *Fire Technology* 19(3) (reproduced in *Fire and Arson Investigator* (1986) 37(2): 42–45).
11 B.V. Ettling (1975) Electrical wiring in structure fires. Northwest Fire and Arson Seminar.
12 R.N. Anderson (1989) Surface analysis of electrical arc residues in fire investigation. *Journal of Forensic Sciences* 34(3): 633–637.
13 D.W. Levinson (1977) Copper metallurgy as a diagnostic tool for analysis of the origin of building fires. *Fire Technology* 13: 211.
14 D.A. Gray, D.D. Drysdale and F.A.S. Lewis (1983) Identification of electrical sources of ignition in fires. *Fire Safety Journal* 6: 147–150.
15 R. Henderson, C. Manning and S. Barnhill (1998) Questions concerning the use of carbon content to identify "Cause" vs "Result" beads in fire investigations. *Fire and Arson Investigator* 48(3): 26–27.
16 D.G. Howitt (1997) The surface analysis of copper arc beads – a critical review. *Journal of Forensic Sciences* 42(4): 608–609.
17 D.G. Howitt (1998) The chemical composition of copper arc beads; a red herring for the fire investigator. *Fire and Arson Investigator* 48(3): 34–39.
18 J.D. Twibell and S.C. Lomas (1995) The examination of fire-damaged electrical switches. *Science & Justice* 35(2): 113–116.
19 B. Béland, C. Roy and M. Tremblay (1983) Copper–aluminium interaction in fire environments. *Fire Technology* 19(1): 22–30.
20 B.V. Ettling (1982) Glowing connections. *Fire Technology* 18(4): 344–349.
21 B.V. Ettling (1981) Arc marks and gouges in wires and heating at gouges. *Fire Technology* 17(1): 61–68.
22 R.N. Anderson (1996) Which came first the arcing or the fire? A review of Auger analysis of electrical arc residues. *Fire and Arson Investigator* 46(3): 38–40.
23 W.E. DeWitt and R.W. Adams (1999). Heat transfer testing of thermal–magnetic circuit breakers. *Journal of Forensic Sciences* 44(2): 314–320.
24 B. Béland (1997) Mechanical behaviour of copper conductors in relation to fire investigation. *Fire and Arson Investigator* 47(4): 8–9.
25 N.J. Carey (2002) personal communication.

The use of laboratory reconstruction in fire investigation

Martin Shipp

Introduction

The complex, random and probabilistic nature of fire means that it is often impossible to fully determine how the fire developed and spread. There will be times when an appropriate and necessary understanding of a particular incident requires a laboratory reconstruction.

Reconstructions can be of particular value to a fire investigation. Those needing a reconstruction can range from the police or scenes of crime officers to insurance loss adjusters to public inquiry members. There are a wide range of questions that may need answering in a major fire investigation, starting from how did the fire start, through to what lessons can be learned from the incident. The objective of any reconstruction must be determined at the start since it may be a 'test', to measure material properties, an 'experiment', to find out what happened or check a specific hypothesis, or a 'demonstration', to illustrate what may have happened.

There are a large range of tests and experiments that can assist the fire investigator. Some tests will be to a defined Standard, others will be ad hoc. They include small-scale materials tests, medium-scale component tests, fire-resistance tests, large-scale assessment of interactions between items or components, full-scale re-creations, and demonstrations for inquiry teams or juries.

Before any reconstruction is attempted, the scale, size and the data required must be agreed. Other issues to consider include the level of realism appropriate, safety, costs and traceability of evidence. Because all fire incidents are different, all reconstructions are different, and so it is almost impossible to fully plan ahead. It is essential to have clear management and good communications on the project, and to keep full records and documentation, but reconstructions will always be expensive and will nearly always be resource limited, both in money and time. While it is sometimes possible to carry out small- to medium-scale ad hoc experiments on site, or at a fire brigade training facility, there are various specialist laboratories able to carry out these various tests and experiments. Full-scale reconstructions, in particular, need the services of a specialist laboratory.

It is always possible to learn lessons from real incidents; to seek to ensure that such events are less likely to re-occur and to identify the 'near miss' features. As well as assisting with the investigation of a particular incident, reconstructions supplement and underpin our existing knowledge base and ongoing research work and provide an essential link to events in the 'real world' to ensure the maximum confidence in future fire safety.

This chapter will discuss these various tests, experiments or demonstrations, and examples of how they have assisted fire investigations.

Why carry out a reconstruction?

There are a number of different organisations or parties that are involved in any fire investigation or inquiry.

Most fires will initially be investigated by fire brigade investigators. Where there has been loss of life or a crime suspected, the investigation will involve the police, scenes of crime officers, forensic scientists, the Coroner or Procurator Fiscal (for fatal accident inquiries in Scotland). If insurance fraud is suspected, or if a major insurance claim is made then the insurers, private investigators, specialist consultants and loss adjusters will be involved.

Once the incident comes to court then it will involve lawyers, the judge, barristers, and jurors. Expert witnesses and scientific advisors may be called upon. Others having an interest, either directly or through their legal representatives, might include the architect, the builder, the building owner, the occupiers, the maintenance engineers, victims, victims' families, the accused and the accused's families. Industrial or work place incidents will involve health and safety inspectors. If a major public inquiry is called, then a number of participants will be helping the inquiry members.

With regards to a particular incident, and in specific relation to any legal action, either criminal or civil, the issues that may be addressed could include the following.

- How did the fire start?
- Who caused it?
- Who is to blame?
- Was it as a result of an act of omission or commission?
- Has a crime been committed?
- Was the fire accidental, deliberate, malicious, arson, or fraudulent arson?
- Can the arsonist be identified?
- Who should pay? (Who else should pay?)
- Was a 'material' or 'construction' significant in the danger or losses from the fire? If so, who can be sued?
- Was the building design significant in the danger or losses from the fire? If so, who can be sued?
- Was the operation of the building, or management procedures, at fault?

Because a fire will destroy its own evidence there is frequently a need for conjecture and opinion based on inadequate facts and consequently there can often be disagreement between experts on what happened.

In addition, and especially in the case of multiple fatality or other high profile incidents, there may be an expectation that lessons will be learned that will go towards improving the law to seek to reduce the risk of similar incidents in the future (usually through regulations, codes or standards). Learning lessons from an incident will often be one of the objectives of a Public Inquiry, a Coroner's Court or a Fatal Accident Inquiry (Scotland). The issues addressed now might include the following.

- What lessons can be learned from this incident?
- Is it likely to happen again, or was it a freak event?
- Can it be prevented from happening again (or the risk significantly reduced)?
- Are there 'political' implications? (especially if government already knew about the particular problem)
- Are the current regulations (codes and standards) adequate?
- Were there unusual or unexpected features of the fire?
- Can it be shown how the fire developed?
- Does the 'public' expect a reconstruction?

In some cases, and especially in high-profile events, other organisations can have an interest. These include the media (particularly television) who often carry out their own investigation, or research bodies who have identified an unusual feature of the incident that demands further study. TV companies are often interested in pursuing a 'pet' theory or a perceived failing in the official processes. Scientists will be interested in using the incident to improve fire statistics and fire science for a whole range of disciplines. The incident represents an uninstrumented experiment from which lessons can be learned about physics, chemistry, engineering, management, biology, pathology, toxicology, human behaviour and psychology, within the context of fire.

All of the different people identified here will want to know the details of the fire but will all have a different knowledge of fire behaviour. Each and any of these (or their legal representatives) may require laboratory tests or a reconstruction to assist in their investigation.

Purpose of the reconstruction

Issues that can be addressed by a reconstruction

Not all of the issues that will be the subject of an investigation, trial or inquiry can be addressed in the laboratory. However, there are a large number of issues that can be addressed, examined or resolved by carrying out a reconstruction that include the following.

- How did the fire start?
- How did a particular item or collection of items burn?
- How or why did the fire-spread?
- How quickly did the fire-spread?
- Could it have spread this way or that?
- Did a particular material contribute?
- How hot did it get?
- How big in area did it get?
- How high did the heat release rate did it get?
- How long did it last?
- What did it look like?
- How smoky was it?
- How did the smoke spread?

- How toxic was it?
- How did the structure respond?
- Did a material behave as would be expected?
- Did a structural element behave as would be expected?
- Did the building design, or a design feature, contribute?
- Were there any unusual features of the fire?
- Did the passive fire protection perform as required? (Passive fire protection includes fire protection coatings, dampers, and doors.)
- Did the active systems perform as required? (Active systems include detectors, alarms and sprinkler systems.)
- Did the fire resisting elements perform as required? (These include compartmentation and load-bearing elements.)

A single test or experiment may not answer all of these and a series of tests may be necessary. But it must be recognised that not all of the questions that need answering can be done so by a reconstruction. There may be key pieces of information missing so that the results of a reconstruction may add no value to the investigation. An example of this sort of problem is if timescales need to be established but the source of ignition has not been identified. In such a case whether the ignition source was smouldering or flaming will have to be assumed and the specific times in the test will be meaningless.

The objectives of the reconstruction

Specifying the objectives of a reconstruction is crucial and no reconstruction should be undertaken unless these are agreed with the client. While a laboratory will assist in the formulation of the objectives, it will be up to the client (i.e. the sponsor or funder of the project) to agree to the specification. It is therefore important for the laboratory to know who the client is, since they may be arranging the reconstruction through agents, such as expert consultants. If the agent has full delegated authority then this must be explicit and referenceable, since the objectives of the reconstruction could be subject to aggressive questioning in Court.

It may be the case that the client is a consortium, or a group of separate parties with a common interest, and a reconstruction can seek to address more than one objective. It is clearly in the best interests of all if a single reconstruction can serve in place of a number of them. Any reconstruction may lead to conclusions not related to the objectives, but (however valid) these must be considered a bonus. There are two ways to identify the objectives. Sometimes the client will come to the laboratory with a problem to solve and the scientist will tell the client whether a reconstruction will resolve the problem. Alternatively, the client will already know what is wanted and how to achieve it. It will sometimes be the case that a reconstruction to resolve a particular issue can, with some modification, be used to resolve other issues, or to provide an additional demonstration of other factors. It also needs to be recognised that the fire science community is quite small and that individuals working on the reconstruction will need to be aware of potential conflicts of interest.

Types of test or reconstruction

It is generally the case that there are four types of laboratory work that can assist a fire investigation.

Forensic laboratory tests

First, there are the 'forensic' scientific examinations, for example, for DNA samples, or the detection of accelerant residues. These are undertaken by forensic science laboratories, either government agencies or private forensic laboratories. This type of laboratory work is not discussed further here.

'Standard' tests

Second, there are 'standard' tests. These are tests, mostly bench-scale, which use a standardised method to determine or check specified material properties. Such tests are usually carried out according to the standard. There are a large number of fire performance standards, British, European, foreign or international, and the most common of these in the UK are given in Appendix A. Sometimes it is appropriate to modify a standard test to meet a specific objective, or to use a standard test method for a material for which that test is not intended. The test will then be considered ad hoc and the test report should state '...carried out using the method of standard test x'. These types of 'standard' test are discussed briefly later, since they can provide a very useful basis for a more elaborate reconstruction.

Experiments

Third are experiments. These are intended to answer specific questions, to gain new knowledge or to test a hypothesis. As discussed earlier, experiments may be required to find out what happened or how quickly it happened or check a specific theory, for example, to see if the fire could have grown to a particular size. Experiments can be of various sizes or scale, from small bench-top experiments to full-size reconstructions. It is the larger of these experiments that is the main subject here.

Demonstrations

Fourth are demonstrations. These are usually medium to large-scale reconstructions in which it is not anticipated that any new knowledge will be derived but are intended only to show what happens to a non-specialist audience, for example, an inquiry team or a jury. Demonstrations will often, fortuitously, lead to some new understanding. Demonstrations are also used for training fire investigators. Reconstruction demonstrations require much of the same planning as large-scale experiments, but usually with less instrumentation. They are also the subject of discussion here.

Designing the reconstruction

Primary issues

As mentioned previously, the first issue for a laboratory is to identify the client or the delegated agent of the client. The objectives of the work must then be identified and agreed. There will nearly always be constraints, of money or time or both, which will impose limits on what can be achieved. Financial, time and other controlling factors must be determined and agreed early on. The extent and reliability of the existing knowledge needs to be defined, and assumptions, both those of the client and those of the laboratory (in order to carry out the reconstruction) must be agreed explicitly. The need for traceability of evidence must be specified if appropriate, and the type of reporting required must be defined.

With any project there needs to be regular discussions with the client, and, where possible, with other participants in the case. It is usually better to consult with other parties early on rather than risk the project falling in Court.

Other planning issues

There are a number of other issues that need to be agreed with the client before the design of the reconstruction commences. These include:

- The use of material from the actual incident. If this is evidence, especially crime scene evidence, how must it be stored, and for how long?
- Storage and physical security of data and other records and documentation. How long will the data and other records from the reconstruction need to be retained? Electronic security of data and other records and documentation. How long will the data and other records from the reconstruction need to be retained and will they be retrievable?
- Contingency and/or liability issues.
- Confidentiality.
- Physical security of the reconstruction (e.g. against media intrusion).
- Ownership of results and 'publication' policy.
- The use of computer models.
- Dealing with the media.

Many of these issues and the reconstruction design will be subject to regular discussion between the client, the client's specialist advisors and the laboratory. It needs to be recognised that there might be times when the client's requests are contrary to the advice of the laboratory. Such requests will need to be documented in the Report.

Type

It is possible to define the type of test or reconstruction only once the objectives have been specified. If 'standard' tests are needed, then these will be conducted according to the protocols that are in place as part of the laboratory accreditation for the particular test. The most commonly used 'Standard' fire tests in the UK are listed in Appendix A.

If an ad hoc test, reconstruction or demonstration is needed then it will be necessary to decide on the method to be adopted, what features or factors are to be demonstrated or investigated, the scale and size of the experiment and the type and quantity of data required. This will often be a balance between the ideal and the practical, determined by the competing factors of cost, time and quality. Sometimes an existing rig can be used for speed and cheapness, but if so, any factors that are then imposed upon the experiment must be carefully considered. Existing instrumentation can be utilised if it is appropriate.

The following types of experiment are available to help the investigation:

Small-scale materials tests

These are bench-scale ad hoc tests based on the standard tests and can provide information on the fire behaviour of a particular material, in isolation, such as a piece of furniture foam, or a wall covering. They can provide information on ignitability, combustibility, spread of flame properties or propensity to self-heating, for example.

Medium-scale component tests

These tests will use an existing test or research facility and involve whole items or components, such as an armchair, or stair carpet on a stairs or a piece of flooring. These tests can include calorimetry (measurement of heat release and smoke and combustion gas production under a, typically, 3 m by 3 m hood), fire resistance tests (on elements of construction up to around 3 m by 3 m, to examine structural stability and heat transmission), and irradiance tests (using a radiant panel, to examine ignition).

Large-scale assessment of interactions between items or components

These experiments can be part-scale, reduced-scale or full-scale reconstructions. Part-scale experiments involve only a selected element of the scene to be reconstructed, for example, the floor, walls and ceiling of one corner of a room (perhaps where the fire is thought to have started). Reduced-scale experiments involve a selected element of the scene to reconstructed, built smaller than the real scene. This can save on cost and time but the implications of the scaling effects, especially on timescales, will need to be carefully considered. Full-scale experiments involve the construction in the laboratory of a large and complete representative portion of the scene. The reconstruction will be fitted out with appropriate instrumentation.

Location

Once it has been decided that a reconstruction is needed, then the location for the experiment must be determined.

Exceptionally, an experiment can be carried out in the actual building of the incident, possibly in a similar location. This is only likely to be possible if the building is not badly damaged by the fire, and has the advantages of essentially identical materials and construction, which can avoid later controversies. Alternatively, a similar building might be located, for example, in a row of houses or on an industrial estate. This option is very seldom available in the UK but has been used successfully in the USA.

Relatively simple and small-scale 'back yard' experiments can be carried out on a nearby open space, perhaps the car park of the affected building, or using the fire brigade training facilities. Such experiments will have very limited instrumentation, if any.

In general, and particularly for major inquires, the facilities of a specialist fire laboratory will be called upon. Such laboratories will either use an existing rig, or construct a special rig, within a specially equipped laboratory or in the open in identified locations.

A number of specialist fire laboratories are listed in Appendix C. Any laboratory selected to carry out the 'Standard' fire tests should be accredited by UKAS (the United Kingdom Accreditation Service), or the equivalent national body, for that particular test.

Realism

The next step in the design process will be to determine how 'realistic' the reconstruction needs to be. This will in part depend upon the objective of the reconstruction, in part upon the importance of the parameters being assessed, and in part on limitations of cost and time. Issues to consider will include the following.

Construction method

Often the least important factor, provided the materials are realistic, the structure or the rig needs to be representative of the actual incident if issues such as collapse could be important. Rigs are typically constructed of timber with ceramic board, or from blockwork.

Materials

The materials, both of the rig construction and any furniture, furnishings or contents, must be carefully considered. Ideally, materials retrieved from the actual fire should be used. Alternatively, a materials audit (a review of the actual fire) needs to be carried out and documented. The materials then selected must be shown to be either identical with those in the incident, or of identical (or adequately similar) properties with regards to fire performance. Alternatively, it must be explicitly shown (and defendable in Court) that the particular materials were not significant in the incident. Materials used for the recreation must be appropriately conditioned (e.g. with regards to humidity).

Dimensions

The dimensions of the recreation will be determined from the objectives of the experiment. A small rig can be used if the early stages of ignition only are being examined. If fire-spread across a room is the issue, then the rig must be that size. As mentioned before, reduced-scale rigs can be built smaller than the real scene (particularly ceiling heights) but the implications of the scaling effects, especially on time scales, will need to be carefully considered. Radiation effects are difficult to scale.

Details

The need to attend to the details will also be determined from the objectives of the experiment. These will include ventilation, ambient conditions, humidity, temperature, wind, condition of the materials (in particular moisture content, but also surface damage, ageing, weathering, and for paints, curing), and physical details such as door cracks and wall lining fixings. As discussed next, these are often the subject of assumptions.

All of these issues need to be documented and included in the Report. It is usually of value for the laboratory scientist who is designing the recreation to see the incident fire scene, but this is not always possible.

Assumptions

In planning and designing a reconstruction, it is essential to identify and record all and any assumptions. This can be crucial in Court, and failure to do so can result in the entire project being thrown out. It is necessary to beware of 'hidden assumptions' that is, where something is 'taken for granted' or 'we always do it that way', for example involving a component or element that no one has thought about. Again, the whole project can fail if a crucial and unconsidered assumption is identified by the other side in Court.

It is necessary for the client and the laboratory scientist to ensure that they have identified and documented all of the assumptions and to critically review the information sources and assumptions that are specified by the client.

Issues to consider include, for example, the layout and dimensions, including details such as door cracks and leakage paths, the materials, the provision of the materials, the ignition source and the environmental conditions. Particular problems arise where specific details are unknown and where a misjudged assumption can have a critical effect on the results of the experiment. As stated earlier, it is usually of value for the laboratory scientist who is designing the recreation to see the incident fire scene, but this is not always possible.

Detailed design

The detailed design of the construction can only commence once the above-mentioned issues have been agreed with the client. These details will then include the following.

- The size and scale of the reconstruction. Will it be small, medium, reduced, large-or full-scale? The size and scale need to be representative or have an established and documented relationship to the real incident.
- The materials of which the rig will be made. There is a need to consider the type, thermal properties, conditioning and curing of these materials. They need to be representative of the real incident.
- The contents, furniture, furnishings or other items, that will go into the rig. Will they need to be specially conditioned or treated to be representative of the real incident?
- The use of actual materials and contents from the incident. If so, how will they be obtained? What condition or age will they be? Will they be traceable? What special treatment or conditioning will be needed?
- The need for any structural elements to have loading, either static and/or dynamic. How will this be determined?
- The ventilation conditions within the reconstruction. These need to be representative of the real incident, but how will this be determined?
- The ambient conditions (humidity, temperature, wind) within the reconstruction. These need to be representative of the real incident, but, again, how will this be determined?
- The duration of the fire. In many cases, the fire will be allowed to burn to completion, but, particularly for large reconstructions, some limits of time or other termination criteria will need to be agreed.
- Audit trails and traceability of evidence. These will need to be determined and agreed.

The ignition source to be used needs to be determined from the incident or else agreed with the client as an assumption. The type and size of the ignition source will significantly affect the timescales of the recreation. Failing any useful alternative, a 'standard' ignition source, such as a standard crib [1] can be used.

In most cases there will always be some residual doubt regarding the details of the design of the reconstruction and the validity or soundness of any assumptions. It will be a matter of judgement as to whether these doubts need be critical to the project, but it needs to be recognised that there will be times when the uncertainties regarding the incident make any reconstruction pointless. It is therefore best if all relevant parties agree with the proposed experiment in all its details. Sometimes an invalid assumption or faulty information from the incident only becomes apparent during or after the reconstruction fire. In such circumstances the possibility of additional reconstructions may need to be considered, but in all cases the interpretation of the results of the reconstruction will require a degree of expert judgement. It is generally the case that it is best to start with small tests and work up towards a large test or experiment.

Reproducibility

An issue which affects all large-scale fire research is that of the number of fire experiments. For fire investigation reconstruction, experiments or demonstrations, a single fire is often sufficient. However, if the effect of a particular parameter is being examined, for example, the quantity of accelerant needed, then a number of fire experiments may be needed, each using a different quantity. As discussed elsewhere, the number of experiments will in part be constrained by cost and time. But often it is possible to carry out smaller-scale trials to examine a range of factors before carrying out a single large-scale reconstruction.

The complex, random and probabilistic nature of fire means that it is often impossible to be fully satisfied that the reconstruction has adequately recreated the events at the actual incident. However, given that there will always be some assumptions inherent in any reconstruction, additional fires will not necessarily resolve the problem.

Instrumentation

The instrumentation in any reconstruction can be a significant cost in equipment, installation and analysis. Clearly sufficient instrumentation will be needed to satisfy the objectives of the project, but where this is limited due to resources the decision process needs to be documented. There are a number of parameters that can be measured during a fire reconstruction which are given here.

Temperature

Temperature is most simply measured using thermocouples. These devices depend upon the voltage produced where two different metals are in contact and need little calibration. They can be made quite cheaply from rolls of the appropriate wire, by welding or silver-soldering, of a thickness and metal type appropriate to the anticipated fire severity. Stainless-steel sheathed thermocouples are more expensive but are usually more robust. A 'cold-junction' is required to give an accurate reading, but this is nowadays almost always incorporated in the data logger. Measurements will be in degrees (Centigrade) and be continuous (subject to logging rate). Specialist help may be needed.

Heat flux

Heat flux (or heat flux density) is measured using off-the-shelf heat flux meters, which are moderately expensive. They are (usually) water-cooled devices, about 25 mm in diameter that measure the temperature difference across a small metal plate. They require careful calibration before use, and need a water supply. Results will usually be in kilowatts per square metre (kW/m^2), and be continuous (subject to logging rate).

Mass loss

Mass loss is a means of estimating heat production. The mass of a selected item can be recorded using load-cells. The load cells must be calibrated and need careful protection from the fire. If the calorific value of the selected item is known, and the combustion efficiency estimated, then the heat production can be derived. However, in real fires the items involved usually comprise a mixture of materials, and so this technique is now seldom used. The following technique is now preferred. The mass measurement can be continuous (subject to logging rate).

Heat release rate and total heat release

These can be measured using oxygen-depletion calorimetry. This involves the use of a large hood and fan, positioned over the reconstruction, to collect all of the fire gasses. The gasses travel down a duct in which instruments measure temperature, velocity (hence mass-flow rate) and oxygen concentration. The mass of oxygen consumed by the fire can be derived and, via now well-established relationships, the heat release rate can be calculated. The measurement will usually be in Watts, kilowatts or, for large fires, Megawatts, and be continuous (subject to logging rate). By integrating the heat release rate over the duration of the fire, the total heat release can be derived (Joules). A number of specialist laboratories have such hoods. The largest in the UK are those at FRS, Garston (9 m by 9 m) and at the University of Ulster. The system must be calibrated prior to the reconstruction.

Gas composition (and toxicity)

A number of different gas species can be measured during a recreation, usually via a tube and pump. Some of these can be measured using online analysers, which can be continuous (subject to logging rate), and include:

- carbon monoxide
- carbon dioxide
- oxygen
- nitrogen dioxide
- nitric oxide
- hydrocarbons.

Others require the gasses to be collected in a flask (called a bubbler) containing an appropriate liquid. These are analysed later, by gas chromatography or mass spectrometry. These measurements will not therefore be continuous over the duration of the fire, but by using a number of bubblers, each running for a selected time period, it is possible to examine the different stages of the fire. Gasses that can be recorded this way include:

- hydrogen chloride
- hydrogen cyanide

- hydrogen fluoride
- hydrogen bromide
- sulphur dioxide.

Gas concentration measurement requires carefully calibrated equipment, and analysis, and is a heavy user of resources.

Smoke density

Smoke density can be measured at selected locations within a reconstruction using optical devices. These comprise a light source and a receiver, rigidly fixed a specified distance apart. The presence of smoke is recorded by a reduction in the receiver signal. The devices need calibration prior to the fire but give a continuous recording. The measurement will be in optical density per meter, or visibility (in m). Smoke detectors (ionisation or optical) can be used to identify the release of smoke, but these will not give a calibrated measurement.

Smoke production

Smoke production can be measured as part of the measurement of heat release (see Heat release rate). The optical density of the smoke within the duct is recorded and can be converted to a total quantity of smoke over the duration of the fire. The measurement will be in m^2. Alternatively, the mass of smoke can be estimated by gravimetric means; smoke particulates are collected on a filter and weighed.

Deflection/stress

Deflection is measured using displacement transducers. Stress can be determined for selected structural elements using strain gauges. These small devices depend upon a change of electrical resistance as the device changes shape.

Visual

Visual recording of the reconstruction will nearly always be carried out, nowadays using video equipment and stills photography. Video records are used to estimate fire growth, flame lengths, fire spread, and the response of items. Infrared photography or video may also be used. Sacrificial video cameras may be used in specific locations. Care needs to be taken with video since the recordings will often be used in Court and a 'burnt-in' time code may be required.

Factors

Planning for the instrumentation of the reconstruction will need to consider a number of factors, which include the following.

- Number of sampling points, number of sensors.
- Location of sensors/sampling points.
- Computer analysis (additional sensors, or different locations, may be needed).

- Duplication of systems.
- Protection of equipment and instruments from the fire.
- Sampling rate(s).
- Data format.
- Data security (including electronic security), long-term data security.
- Accuracy, sensitivity, calibration.
- Data logging, processing and analysis.
- Observations locations.
- Video cameras, stills photography; format, number.
- Physical security of recordings (tapes, discs, etc.).

Safety

Fire experiments, by their very nature, involve some dangers to personnel. Most specialist fire laboratories will have well-established safety procedures that will be called upon. These dangers will include flames and heat, particulate smoke (which can obscure vision), and combustion gases (which can be irritant and/or toxic). In addition, within the test rig, there can be risks from chemicals, toxic materials, electricity, structural collapse and explosion. Working within the laboratory entails risks from impact, drops, trips and falls, vehicle movements, etc. It is also necessary to be aware of, and mitigate, any environmental risks, from toxic materials, smoke or polluted water run-off.

The test rig should be designed to minimise risks and be able to be constructed safely. Overhangs and trip hazards should be avoided, but may be essential if part of the re-created layout. Electrical safety must be considered, especially, as is usual, water will be used to extinguish the fire on completion or if needed for safety. Bunds and reservoirs should be provided for fuel spillage and/or water run-off. Provision must be arranged for the safe disposal of liquid and material waste. The potential toxicity of these materials after a fire may require the services of specialist waste contractors.

A safety plan needs to be developed early in the project and integral to the experiment design. The safety plan will need to include a risk analysis and a description of how the risks are to be managed. The plan will need to consider the safety of observers and visitors, communications, access/egress, safety routes, places of safety, evacuation criteria and procedures, first-aid and first-aid rooms. Again, where observers and visitors will be present, the test rig will need to be cordoned with barriers and good signage. Sirens, klaxons or a public address system may be needed to communicate with large numbers of visitors. If large quantities of smoke are anticipated then direct viewing of the fire may not be acceptable and a video link to a safe observation room may be needed.

A well-defined safety team will be needed, with formal responsibilities and powers. There must be clearly identified tasks and roles. For larger experimental fires, the local fire service may be employed for firefighting and clear lines of communication and command will be established before the test date, with agreed safety protocols. Similarly, a first-aider may be needed in attendance.

On the day of the test, the Safety Officer will need to provide all participants with the safety protocol and give a safety briefing, describing the command structure and procedures. Agreeing to abide by the instructions of the Safety Officer will be a condition of attendance.

Termination criteria (for the fire) must be set.

Although the experiment should be designed to minimise risks, appropriate safety equipment must be available to the personnel. These may include overalls, safety boots, hard hats, safety glasses, gloves, gas monitors (usually CO), additional lighting and if appropriate, respirators or breathing apparatus. First-aid facilities should be identified.

The provision of appropriate safety arrangements for a large-scale fire experiment can be a significant added cost to the project. Despite the best planning, there will always be some residual risks and these must be proportionate to the importance of the project.

Conducting the reconstruction

Personnel

The reconstruction will involve a large number of people and tasks. These might include:

- Officer-in charge
- Experimental staff
- Instrumentation (logging) staff
- Technical observers
- Safety Officer
- First-aider
- Steward (to look after visitors)
- Laboratory Audio Visual Crew
- Client's Audio Visual Crew
- Fire Brigade
- Clients
- Guests
- Catering staff.

Guests at the recreation will be at the discretion of the client and might include the judge, jury, barristers, and experts from the various teams. Special arrangements are sometimes needed where a large number of guests are expected, including catering and washrooms.

Schedule

The schedule for any reconstruction may start weeks before the actual fire experiments. As well as resolving and organising the various design and planning issues as discussed before there will be a number of other tasks that will need to be programmed. These include:

- co-ordination with other users of the laboratory;
- organising or subcontracting construction of the rig;
- purchase of consumables;
- storage of materials, especially fuels;
- identifying and resolving any environmental issues;
- rig construction;
- supplementary construction and installation (e.g. bunds, ventilation systems);

- obtaining materials, furniture, etc. and conditioning;
- calibration of instruments;
- fitting instruments and data logging;
- fitting of ignition system and back-ups;
- commissioning rig, equipment and instruments;
- liaison with fire brigade;
- briefing team members;
- planning for clearing up after the fire, hiring skips, etc. and contracting specialist waste removal companies.

For the planned day of the fire experiment all attendees need to be provided with a programme in advance. The fire should be started only when everything is ready, so all attendees need to be made aware that precise scheduling is not possible. Once alight, it is seldom possible to deviate from the agreed schedule. All visitors need to be made aware that there can be considerable waiting time, often in an, initially, very cold laboratory.

Clients, visitors and guests will normally wish to inspect the reconstruction and care needs to be taken since many will not have previous experience of fire experiments. 'Guided tours' may need to be arranged. As well as safety considerations, there is a need to avoid damage to instrumentation.

All those attending, including laboratory staff, should be given a safety briefing, which will include identification of escape routes. Clocks and stopwatches should be synchronised.

Once observers are in place, a countdown can commence and logging, recording and video equipment can be started. At the planned time-point the fire can be started, either manually (e.g. with a lighted taper), or by an electrical device. Ignition is not always successful and a second start can be required. The time of restart will be recorded.

If the termination criteria are achieved, the fire will be extinguished. Alternatively, it will be left to burn out. Loggers and recorders may be switched off or may be left to run on. Clients and guests will wish to review the aftermath of the reconstruction. However, it is seldom possible to re-examine the reconstruction to assess damage until the rig has cooled and the smoke dispersed, which may take from an hour to a day. Safety protocols must be followed.

Until deemed otherwise, the site should be treated like a crime scene. Photographs will be taken and, where necessary, samples will be logged and removed. Instruments will be extracted.

The remains of the reconstruction will then need to be disposed of in accordance with the previously agreed procedures.

Results will then need to be processed and analysed. This may take some days (or weeks) depending upon the complexity of the experiment.

Reporting

The Report of the reconstruction, including photographs and supplemented by any video recording, will be the output of the project, and is very likely to be presented in Court. It therefore needs to be written with this in mind. This document will usually be written by the laboratory scientist and, subject to the requirements agreed with the client, may be a report of the results, or may include deductions and conclusions. Contrary to scientific practise, it is often better for the report to have a single author, even when a project team has been involved, since any of the named authors may be called into Court.

In general, the report should conform to normal good practice for expert evidence and will need to include:

- An introduction, including identification of the Client
- Background
- Instructions from the Client and purpose
- Appropriate declarations and statements
- Prior information
- Assumptions
- The reconstruction design, including geometry, materials, and issues of traceability
- Ignition source
- Instrumentation
- Results and observations
- Analysis
- The use of computer models, results, etc.
- Discussion, including deductions, if required
- Conclusions
- Acknowledgements, including the project team
- References
- Photographs.

It is essential that the laboratory have a quality and checking protocol to minimise the risk of typographical or other errors.

Often the findings from a reconstruction have wider scientific value. The results of the project will be owned by the Client and publication of the Report will be at the discretion of the Client. Courts of Inquiry will usually wish to publish. Once a report has been presented to the Court it is effectively in the public domain, but the cost of publication will be an issue. The laboratory scientist may be required to present the Report in Court. The Report can then be supplemented with the video record of the reconstruction.

Costs

It will have been seen from the discussions here that full-scale reconstructions do not come cheap. Bench-scale and 'standard' tests are much cheaper, but often can only provide some of the information needed in an investigation. Clients will naturally be seeking 'value for money', but an over-simplified experiment can be worse than useless since it can undermine the credibility of the Client.

The costs of the project will include:

- design, customer meetings, and staff;
- primary structure, construction, laboratory space (floor space rental) and facilities;
- storage and conditioning facilities;
- materials, fuel, the contents and furnishings;

- equipment, instruments, calibration, means to protect the instruments;
- audiovisual (stills camera and video cameras);
- the Fire Brigade and other safety requirements;
- clearing up afterwards, hire of skips and/or specialist waste removal companies;
- storage of evidence, storage of data (especially for an extended duration or under special conditions);
- travel and catering;
- data analysis and producing report;
- video editing;
- contingency and liability considerations;
- environmental protection issues;
- confidentiality, secrecy, physical security;
- some cost allowance for the laboratory scientist to give evidence in court. However, this is often treated as a separate issue.

It follows that all of these issues must be considered and agreed with the client at the start of the project.

Case studies

The following selected case studies are intended to illustrate the value of recreations, at the different scales as discussed before.

Stardust Disco (Dublin), February 1981

Following a series of small-scale tests, a full-scale reconstruction was carried out of the Stardust Disco (Dublin) fire [2]. This required a very large rig, clad with carpet tiles and fitted with tiered seating of the same type as in the incident. The fire was started at the rear of the set-up. The test demonstrated the speed and severity of the ensuing flashover fire.

Windsor Castle, November 1992

A medium-scale test was carried out to examine the fire development of the curtain in the Windsor Castle fire [3]. The fire was understood to have been caused by a hot halogen spotlight. Curtain of a type similar to that in the fire was set up on a gantry in the laboratory. The test was to show how quickly, and how, the fire would spread up the curtain.

Dumfries House Fire, February 1995

This project required an assessment of the contribution from the carpet to a fatal fire in domestic premises. An existing test room was used and fitted with carpet identical to that in the fire, and an armchair from the incident house. The fire was started on the armchair. The test showed how the carpet became involved in the fire under the heat from the chair [4].

Computer models for the fire investigator

Computer models are increasingly appearing in Court and are being used as an adjunct or alternative to physical reconstructions. The fire investigator now needs to have familiarity with computer models, both to assist their own work, but also so that he/she might respond effectively to the results from models presented by other parties.

Currently, such models include the following.

- 'Calculator' packages, which link established equations.
- Zone models, to calculate smoke and gas movement in simplified conditions.
- Computational Fluid Dynamics (CFD) (Field models) to calculate smoke and gas movement in complex conditions. Fire-specific CFD models are available.
- Structural models, to calculate the loads and distribution of loads in the building structure when selected elements are heated by the fire. Some such models use finite element analytical techniques.
- Radiation models, to calculate the heat flux from a flame to a complex geometry.
- Sprinkler models, to estimate the response of a sprinkler head.
- Risk models, to estimate the numbers of injuries or deaths in a building under a very large number of different potential conditions. Some such models use 'Monte Carlo' sampling methods.
- Evacuation or egress models, to calculate times to empty a building, or the distributions of people leaving a building. Current models can include a variety of behavioural responses.

Some investigators use virtual reality visualisations, but by themselves these are only a means to providing a three-dimensional, moving and time-sequenced diagram, 'drawn' by the scientist, and extreme care must be taken in their use since they do not 'predict' an outcome. Such models are, however, now being linked to some of the above-mentioned computational models to assist in the presentation of the outputs. In these cases it must be clear what is predicted by the model, and what is drawn.

Very few fire investigators currently use computer models in their work and CFD models in particular need skilled users. A few models have been specifically developed by forensic scientists, fire investigators or fire investigation institutes, mostly outside of the UK.

While computer models are being developed for a very wide range of fire-related scientific issues, those of concern here are the models used to calculate and predict the spread of heat and smoke and are of the most value to the fire investigator since they can be used to assist the investigator to determine the speed of smoke spread or the temperatures as a result of a particular fire.

Fire growth and spread

Before a full theoretical model of fire can be constructed, a proper understanding is required of the complex and interacting mechanisms involved. The very strong coupling between the fire source and its enclosing structure makes fire a particularly complex problem to analyse. However, theoretical models of varying levels of complexity have now been developed which, with the aid of the computer, can be brought together to begin to describe realistically these phenomena. Such theoretical models have now advanced to the point where they are used with some confidence to predict how the hot smoke and gases produced by a proscribed fire source will be dispersed throughout a building.

A complete treatment of the growth and spread of flame under realistic conditions, where complex arrangements of furnishings and building linings are involved, is not yet within the compass of their capabilities. But simplified analyses do permit an assessment of fire growth and spread to be made. To appreciate the types of computer model that are currently available it is necessary to understand at least broadly, differences in approach adopted by the various models.

Theoretical computer models that simulate the heat and mass transfer processes associated with a compartment fire, fall broadly into one of two categories. These are commonly referred to as models of either the zone or CFD type. The essential difference is in the way they treat the movement of the products of combustion within the building envelope and their respective reliance on experimental information.

Both types of model can predict gas temperatures resulting from a specified fire in a domestic-size room and both give broadly similar results. The field model predictions are much more detailed but, more importantly, are able to show detail such as a clear deflection of the plume caused by the fire-induced wind flowing through the doorway opening. The zonal model does not do this, unless it has been assumed a priori.

It is also necessary to differentiate between deterministic models and probabilistic models. The models described here are deterministic. Probabilistic models tend to have much simpler calculations but a very large number of simulations are conducted.

Zone modelling

Computer zone models are closely related to well-established traditional methods for the treatment of smoke movement, which were initiated before the widespread availability of the modern computer. They rely on a number of simplifying assumptions concerning the physics of smoke movement suggested by experimental observation of fires in compartments.

The availability of the modern computer has allowed this simplified approach to develop further to allow examination of growing fires and to include many more of the relevant influences such as heat loss to the surrounding structure and radiant ignition of remote objects.

In addition to permitting full enclosure fire simulations, the computer now offers easy access to some of the more simple semi-empirical zonal relationships that can be found in engineering guides and handbooks. Because of the ease with which these calculations can be repeated many times over, they are particularly valuable for undertaking preliminary scoping studies. But the zonal philosophy has limited general use because it is dependent on assumptions of how the products of combustion will behave.

A zone model may be specific to a particular problem and cannot be used for other applications.

CFD modelling

CFD modelling represents a break with the traditional approach used in zone modelling. Making essentially no assumptions about the physics of smoke movement, it exploits the techniques of CFD to deduce how, and at what rate, smoke would fill an enclosure. It does this by avoiding resort, as far as is currently possible, to experimental correlations and returning to first principles to solve the basic laws of physics for fluid flow. As a consequence, this type of model is of universal applicability. For this approach, the computer is the enabling technology; without it, the technique could not have developed because it involves the solution of mathematical equations for every step forward that the simulation makes. This is why the zonal type of approach, conceived before this capability had been realised, had of necessity to resort to simplifying assumptions.

The difference between the two types of model is usually only of secondary importance when applied to small compartments but can become very significant in large ones. This is because the zonal method assumes that smoke fills an enclosure from the ceiling down. Fires in tall enclosures, such as atria, may not necessarily behave in this way if, for example, ambient conditions ensure a sizeable temperature gradient between floor and ceiling before the occurrence of fire. In large area enclosures, smoke may not remain in a buoyant layer; it may cool and mix with the lower ambient air before it can reach the outer bounds of the enclosure. It may also be necessary to simulate the conditions before the fire started.

Simulations using field models are able to predict, without prior judgement, the behaviour of smoke flow from a specified fire and to permit smoke control strategies for such circumstances to be assessed using the computer. Like their zonal counterparts, these models allow comparisons to be made between the developing hazard and the time available for safe escape of the occupants.

Models of this kind, being rather more complex and demanding of computer power than their zonal counterparts, are still somewhat restricted to specialist users. Change is, however, taking place as computer power becomes progressively cheaper and simulation models are restructured to admit their use by fire safety practitioners.

Both types of computer model can provide information on when automatic detectors or sprinklers are likely to operate and on the degree of heating of structural elements, flammable contents and of occupants effecting their escape.

The decision whether to employ zone or field modelling depends upon the particular application. Zone modelling will be cheaper but the assumption on which it is based may not be applicable when used, for example, in the design of smoke control for compartments of large volume. Similarly, when designing for early fire detection in compartments of any size, where the energy released by the fire is still comparable with heat associated with ambient air movements, field modelling becomes essential.

What theoretical modelling is not yet able to provide, in any rigorous sense, is a prediction of fire size or growth rate for the generality of flammable materials likely to be found in buildings.

Instead, rates of release of fuel volatiles, steady or growing with time, must be provided as input data to the models. These can be determined for a particular fire load using experimental data, fire statistics, expert judgement or a combination of all three.

Method

CFD (field) models involve setting up a large number of cells to represent the space to be modelled. These cells will form a grid; in some models the geometry will be fitted onto an existing grid, in others the grid will be constructed specifically around the geometry. The computer then solves the equations that determine the flow, velocity and temperature, of the gasses entering and leaving each cell, and, since the flow out of one cell will be the flow in to the next, a very large number of iterations are needed to derive a solution.

Most fire simulation models have been developed so that they can be used as part of the building design process. However, they may also be used to seek to reproduce real fire incidents (e.g. the Kings Cross fire [5,6]). This is done systematically when validating the models. Experiments are designed and instrumented to demonstrate that particular features of the model function correctly and provide an indication of the models' accuracy. Material properties in the experiment are known and most importantly the fire heat release rate can be measured for use in the models. If the models are used to investigate real fires then much of the key input data has to be deduced from the scene of the incident and from reports of witnesses. Estimation of the fire's heat release rate may have to be drawn on a reconstruction of the fire or data from experimental fires using similar materials and configurations (e.g. the NIST worldwide website includes a number of heat release rate curves for burning items such as pieces of furniture).

To investigate a real fire incident the fire engineer needs to select an appropriate model or models, know enough about the design and material properties, and estimate the relevant heat release rate curve. This may result in having to run a number of simulations to resolve 'What if...?' questions by comparing the simulation results with known events during the fire. Selection of an 'appropriate' model requires knowledge of the model's assumptions and limitations as well as its functionality.

Models used

Computer models that have been used to examine fires in buildings, post hoc, include the following.

- JASMINE [7] (Analysis of Smoke Movement In Enclosures) a CFD developed for fire applications;
- CFAST [8] a multi-compartment zone model;
- CRISP [9] (Computation of Risk Indices by Simulation Procedures) a multi-compartment zone model including human behaviour;
- ASKFRS [10];
- FPEtool [11].

There are a large number of other models that are available to the fire engineers. Models that can be downloaded from the NIST website [12] currently include the following.

- ALOFT-FTTM – A Large Outdoor Fire plume Trajectory model – Flat Terrain;
- ASCOS – Analysis of Smoke Control Systems;
- ASET-B – Available Safe Egress Time – BASIC;
- ASMET – Atria Smoke Management Engineering Tools;
- BREAK1 – Berkeley Algorithm for Breaking Window Glass in a Compartment Fire;
- CCFM – Consolidated Compartment Fire Model version VENTS;
- CFAST – Consolidated Fire and Smoke Transport Model;
- DETACT-QS – Detector Actuation – Quasi Steady;
- DETACT-T2 – Detector Actuation – Time squared;

- ELVAC – Elevator Evacuation;
- FASTLite – A collection of procedures which provide engineering calculations of various fire phenomena;
- FIRDEMND – Handheld Hosestream Suppression Model;
- FIRST – FIRe Simulation Technique;
- FPETool – Fire Protection Engineering Tools (equations and fire simulation scenarios);
- Jet – a model for the prediction of detector activation and gas temperature in the presence of a smoke layer;
- LAVENT – response of sprinkler links in compartment fires with curtains and ceiling vents;
- NIST Fire Dynamics Simulator and Smokeview – The NIST Fire Dynamics Simulator predicts smoke and/or air flow movement caused by fire, wind, ventilation systems, etc. Smokeview visualizes the predictions generated by NIST FDS.

When freely available models are used, the user must demonstrate that they are competent in its use. These types of model cannot be used as 'black box' packages.

Selection of the appropriate model

Several factors influence the choice of models to be used. These include range of features, level of detail and degree of validation. In practice other factors such as availability, reliability, cost and user familiarity are also considered. The approach often used is to begin with simple models to gain an overview of the problem and then refine the data input and use a more complex and detailed model to obtain more accurate (or complete) results [13].

Using a combination of modelling techniques, starting with the simpler methods and progressing to more complex ones, can efficiently guide the investigator to a realistic scenario for the incident. However, for this the following are required.

- Awareness of each model's functionality and limitations so that the results from different models can be understood, for example, layer depths and temperatures from zone models compared with temperature distributions from CFD models.
- The construction of a realistic heat release rate curve drawing on a number of sources including:
    - sparse experimental data and small-scale tests;
    - witness evidence (this may be distorted due to poor recollection of events, stress or intent);
    - engineering judgement (e.g. selection of a fire growth curve).
- Material properties (density, specific heat capacity, thermal conductivity, etc.) of building materials under fire conditions.
- Determination of the pre-fire conditions, such as:
    - materials present;
    - position of doors and windows (open or closed);
    - likely locations of ignition source(s);
    - function of heating, ventilation and air conditioning and building management systems;
    - ambient conditions, including wind pressures.

This may require following a number of possible sequences of events using the results from the models to justify or eliminate different combinations. Of particular concern to the fire investigator are the limitations in the use of these models which are given next.

Heat release rate

As mentioned before, there are currently no models available that can calculate the heat release rate of an arbitrary fuel. The heat release rate, either steady state, or time varying, must be input by the user. Since this parameter is often the one that is of greatest interest to the fire investigator, it will be appreciated that this imposes significant limitations on the use of these models. In practice, heat release rates are assumed by the user and estimated from experimental data. There is now a large body of heat release rates for a wide variety of items.

Some of the information required may be obtained from the fire investigation. However, unless this is done specifically with the intention of modelling the fire, the accuracy and detail may be of limited value. For example, the UK fire reporting form includes entries for the fire area when first discovered and fire area when first firefighter arrived. In some situations, it may be possible to give an accurate entry (e.g. one pallet of boxes burning when discovered, three when first firefighter arrived). In other cases, an initial observation of the fire may be more tenuous, such as an initial fire area of 1 m^2. The modeller needs to make a judgement on the accuracy of each data item.

Items in a compartment in a real incident may not correspond to its expected use. While it may be reasonable to adopt the NFPA 't^2' heat release rate curves [13] for a particular occupancy for a design calculation this may not always be appropriate for investigation of a real incident.

At the present time, there are very few models that can be worked in reverse, so that the end-state, as found at the incident can be input to the model and then the model is able to compute the starting conditions (such as heat-release rate).

Other starting conditions

As mentioned previously, there will almost always be a need to make assumptions regarding the starting conditions for any simulation, such as ambient temperature, wind direction and velocity, state of doors and other ventilation. While many of these variables can be assessed by carrying out a number of runs with different conditions, it is important that they be carefully defined.

Validity

The validation history of the model needs to be considered in relation to the actual scenario being modelled, and the investigator needs to be satisfied that the model is sufficiently well proven for the application. Many models suffer from a lack of validation for the application in the field of scientific fire investigation. There is a need for further study and a number of laboratories are pursuing this issue [14,15].

Other areas of concern include:

- sensitivity to inputs;
- risk of misuse;

- sensitivity to data assumptions;
- context of applicability;
- the skill of the user.

Interpretation of results

It may be relatively easy to obtain a result from these models, but difficult to interpret or to determine its accuracy. Often expert interpretation is needed and the complexities of the modelling may be difficult to present in a way that can be properly understood by non-specialists in the Courtroom. Detailed numerical outputs can be hard to comprehend, but over-simplified presentation can be misleading.

It also needs to be appreciated that some of the more complex models have significant resource requirements (time and money consuming), in part because of the size of computer needed, and in part because of the data input effort. Thus the needed number of 'runs', appropriate to give a valid picture of the event, may be scrimped.

The use of computer models in fire investigation has real potential once many of the practical limitations are overcome and there is now an increasing interest in the use of computer models to assist fire investigation [14–17]. The use of CFD modelling to support the King's Cross inquiry [5,6] highlighted the potential benefits for both the fire investigation and the wider fire science community. Many of the issues discussed earlier with regards to design, assumptions and reporting apply equally to the use of computer models in Court. Models can often give an erroneous impression of accuracy which can mislead an inexpert jury, especially where there are colourful dynamics displayed.

Learning lessons from fires

Fire safety systems differ from nearly every other engineering system in a building since any faults or failures in design, implementation or maintenance will only become apparent during the very emergency for which they are required. With the move into the new era of complex fire safety engineering it is becoming increasingly important that information from real fires is fed back into the fire science knowledge base.

For any building project, the fire engineer must demonstrate compliance with the Building Regulations [18], but these specify only 'functional' requirements. Buildings designed to satisfy the recommendations of Approved Document B (AD B) [19] are usually assumed to satisfy the Regulations, and such designs can be assessed and approved by Building Control Officers by reference to AD B. However, buildings designed to the recommendations of well-established and/or widely agreed fire safety engineering codes, such as BS 7974, Code of Practice on the Application of Fire Safety Engineering Principles to the Design of Buildings [20], may also satisfy the regulations. In due course the planned BS 9999, Code of Practice for fire safety in the design, construction and use of buildings [21], is intended to fulfil an intermediate role between AD B and BS 7974.

Other industries, such as the rail industry, adopt a similar approach, where an 'engineered' solution needs to demonstrate equivalent safety to the prescriptive Code [22].

It is important to be able to constantly re-evaluate the knowledge base that underpins fire engineering design. An effectively engineered fire safety system for life and property protection requires a co-ordinated interaction of a number of sub-systems which include the initiation and development of fire, spread of smoke and toxic gases, fire-spread, detection and alarm, suppression, fire service intervention and evacuation. Different fire protection measures are required at different stages of fire development, and this depends upon whether the fire safety system is designed primarily for life safety or property protection. Further complexity arises due to the fact that the timescales for the response of active fire protection measures such as detectors and sprinklers are different from the response time of occupants during evacuation or the structural response time for structural integrity. Assessments of alternative design strategies may depend upon a risk-assessment, where the probability of a particular cascade of events, and its outcome (as measured by deaths, injuries and/or damage), is compared with that of an alternative. 'Risk-assessments' have always been implicit in any fire safety design (or any other safety design). But now such risk-assessment techniques are becoming more quantitative and lead to inevitable concerns (such as 'acceptable' losses). The reliability of the data used in these risk-based approaches has become critical.

Similarly, computer modelling tools require the input of data and some of this is at present conjectural, or subject to specified assumptions (such as burning rate). There is a need to be able to use fire simulation tools with confidence, where there is a pedigree of validation and a proper understanding of the assumptions, limitations and interpretation of the results.

Examples of the issues that need input from real incidents are listed here.

- *Fire load* What are typical (or design case) fire loads in buildings? Are the current Code assumptions sound (or still sound)?
- *Escape time* How long do real people take to escape? How long is spent thinking, how long travelling, how long going the 'wrong' way?
- *Escape route choice* Do people ever use emergency exits?
- *Detection* Do detectors work? Do they operate alarms as assumed? Is the information used as quickly as is assumed?
- *Ignition sources* Are the assumptions regarding sources of ignition sound? Are there some not considered, are there some that really never happen?
- *Fire-spread* Does fire spread in the way assumed? Is it as fast, or as slow, as assumed?
- *Material and structural properties* Do the test methods give a satisfactory indication of the performance of materials or structures in real fires?
- *Flashover* Are the assumptions regarding the processes that lead to flashover valid?
- *Smouldering* What are the mechanisms by which a smouldering fire will go to flaming? How often do burning cigarettes start flaming fires?
- *Visibility* How do people react to smoke in a real emergency? How are decisions made?
- *Tenability* Experiments on people are notoriously difficult. Can the conditions experienced by victims be quantified and correlated with their injuries?
- *Signage* Do people ever look for emergency signs in an emergency? Do they ever obey them?
- *Weather* How and how often do adverse (or benign) weather conditions affect the outcome of a fire?
- *X – the unknown* Are there factors that influence the outcome of a fire that are completely missing from current design analyses?

- *Myths and legends*   Are there factors that are carefully included in design analyses which have absolutely no influence on the outcome of a fire?
- *Fire risk*   How probable (how frequent) are the various events that make up a fire incident? This is possibly the single biggest issue that requires a substantial input from fire investigations.

A few examples of how information from fire investigations have fed back into fire engineering include:

- Woolworths (Manchester), May 1979: this incident showed that stacked goods could shield a significant part of the fire load from the sprinkler spray.
- Stardust Disco (Dublin), February 1981: this fire demonstrated the differences in fire behaviour of material used vertically from the same material used horizontally.
- King's Cross (London), November 1987: the investigation into this incident led to the identification of the 'trench effect'.
- Four Seasons Hotel (Aviemore), January 1995: this fire identified the effect of poor cavity fire stopping and the potential effects and impact of adverse weather conditions.
- The Channel Tunnel, November 1996: this event demonstrated the importance of the management of fire safety.
- Ladbroke Grove (London), October 1999: The post-crash flashed-over fire resulted from the 'wick' effect when two components, each not easily ignited, were in combination.

Other, less well known, incidents have highlighted the importance of design components such as nail plates in roof construction, plastic eaves, weather-proof cladding, sandwich panels, candles, television sets and video cassettes.

As fire safety engineering develops, the effectiveness, performance and reliability of the passive, active and procedural fire safety systems being introduced into buildings becomes more critical if lives are to be adequately protected from a fire. The data, assumptions, methodologies and models that go into fire safety engineering designs must be well founded and reflect what happens in the 'real' world.

Conclusion

Because all fire incidents are different, it follows that all reconstructions are different and it is almost impossible to plan ahead. Nevertheless, some knowledge of what is possible and what is valuable, can help planning for a re-creation early in an investigation.

Lines of responsibility must be defined; there is a need for clear management and good communications from start, and, for the laboratory, it is essential to identify the clients, or the client's agents, to specify the client's needs and hence to define the objectives of the re-creation. Reconstructions are expensive and will nearly always be resource limited, so it is necessary to identify any constraints (mostly time or money) and existing knowledge of the incident. Any assumptions that are needed must be agreed and documented.

The Report of the re-creation must clearly state any assumptions and needs to be written to ensure that it will be understood by a non-technical audience and stand up in Court under hostile questioning.

Acknowledgements

This chapter is contributed by permission of the Chief Executive of the Building Research Establishment Ltd. I would like to thank Richard Chitty for his help with the issue of computer modelling.

References

1   BS 5852: Part 2: 1982; Fire tests for furniture, Methods of test for the ignitability of upholstered composites for seating by flaming sources. BSI.
2   Report of the Tribunal of Inquiry on the fire at the Stardust, Artane, Dublin on the 14th February, 1981, Dublin, 1982.
3   Fire Protection Measures for the Royal Palaces (1993) A report by Sir Alan Bailey KCB, Department of National Heritage, London HMSO, May.
4   Andrew Russell (1996) Concern over carpeting after fatal house fire, *Fire*, March.
5   Investigation into the King's Cross Underground Fire, Department of Transport, HMSO, 1988.
6   G. Cox, R. Chitty and S. Kumar (1989) Fire modelling and the King's Cross fire investigation. *Fire Safety Journal* **15**: 103–106. (See also K. Moodie, and S.F. Jagger (1987) Fire at King's Cross Underground Station, 18 September 1987, Health and Safety Executive, 1987, and The King's Cross Underground fire: fire dynamics and the organisation of safety, Seminar organised by the Institution of Mechanical Engineers, 1 June, 1989.)
7   G. Cox and S. Kumar (1987) Field modelling of fire in forced ventilated enclosures, *Combustion Science and Technology*, **5**, 7–23.
8   R.D. Peacock, G.P. Forney, P.A. Reneke, R.M. Portier and W.W. Jones (1993) 'CFAST, the Consolidated model of Fire growth and Smoke Transport'. NIST Technical Note 1299.
9   J.N. Fraser-Mitchell (1998) Modelling human behaviour within the fire risk assessment tool 'CRISP'. Human Behaviour in Fire – Proceedings of the First International Symposium, pp. 447–457, University of Ulster.
10 R. Chitty and G. Cox (1988) ASKFRS: an interactive computer program for conducting fire engineering estimations, AP46, BRE.
11 H.E. Nelson (1990) FPETOOL – Fire protection tools for hazard estimation, NISTIR 4380, NIST, Gaithersburg, MD, USA.
12 NIST web site: http://www.bfrl.nist.gov/
13 *The SFPE Handbook of Fire Protection Engineering* (2002) third edition, National Fire Protection Association and the Society of Fire Protection Engineers, USA.
14 R. Chitty and J. Foster (2001) Application of computer modelling to real fire incidents, Interflam 2001, Edinburgh 17–19 September.
15 G. Cox (1987) Simulating fires in buildings by computer – the state of the art, *Journal of the Forensic Science Society*, 27(3), 23 pp.
16 H.E. Nelson (1994) Fire growth analysis of the fire of March 20, 1990, Pulaski Building, 20 Massachusetts Avenue, N.W., Washington, DC, NISTIR 4489; 51 pp. June.
17 Z. Yan and G. Holmstedt (2001) Investigation of the dance hall fire in Gothenburg, October 1998 – a comparison between human observations and CFD simulation, Interflam 2001, Edinburgh 17–19 September.
18 The Building Regulations 1991, S.I. 1991 No. 2768 plus amendments.
19 The Building Regulations 1991, Approved Document B Fire Safety, HMSO 2000.
20 BS 7974: 2001, Code of Practice on the Application of Fire Safety Engineering Principles to the Design of Buildings, BSI.

21 BS 9999, Code of Practice for fire safety in the design, construction and use of Buildings, at the time of writing limited to private circulation within BSI.

22 British Standard Code of practice for fire precautions in the design and construction of railway passenger carrying trains. BS 6853: 1999.

Appendix A: The 'standard' fire tests

Fire resistance

Fire resistance is a measure of the ability of a construction element to stay up and prevent the passage of fire, heat or smoke for a specified period, usually long enough for occupants to escape. The test methods are given in BS 476-Part 20, etc. and presume that a quite severe (post-flashover) fire has developed.

Reaction to fire

Reaction to fire tests assess a number of properties and materials. Ignitability assesses whether a material is likely to catch fire. Combustibility assesses whether a material will burn and add to a fire when subjected to an existing fire. Spread of flame assesses whether the fire will spread over the surface of the material (especially wall linings).

The following Standard tests are those for building products:

BS 476-3:1975   Fire tests on building materials and structures. External fire exposure roof test.

BS 476-4:1970   Fire tests on building materials and structures. Non-combustibility test for materials.

BS 476-6:1989   Fire tests on building materials and structures. Method of test for fire propagation for products.

BS 476-7:1997   Fire tests on building materials and structures. Method of test to determine the classification of the surface spread of flame of products.

BS 476-10:1983   Fire tests on building materials and structures. Guide to the principles and application of fire testing.

BS 476-11:1982   Fire tests on building materials and structures. Method for assessing the heat emission from building materials.

BS 476-12:1991   Fire tests on building materials and structures. Method of test for ignitability of products by direct flame impingement.

BS 476-13:1987, ISO 5657:1986   Fire tests on building materials and structures. Method of measuring the ignitability of products subjected to thermal irradiance.

BS 476-15:1993, ISO 5660-1:1993   Fire tests on building materials and structures. Method for measuring the rate of heat release of products.

BS 476-20:1987   Fire tests on building materials and structures. Method for determination of the fire resistance of elements of construction (general principles).

BS 476-21:1987   Fire tests on building materials and structures. Methods for determination of the fire resistance of loadbearing elements of construction.

BS 476-22:1987   Fire tests on building materials and structures. Methods for determination of the fire resistance of non-loadbearing elements of construction.

BS 476-23:1987   Fire tests on building materials and structures. Methods for determination of the contribution of components to the fire resistance of a structure.

BS 476-24:1987, ISO 6944:1985   Fire tests on building materials and structures. Method for determination of the fire resistance of ventilation ducts.

BS 476-31.1:1983   Fire tests on building materials and structures. Methods for measuring smoke penetration through doorsets and shutter assemblies. Method of measurement under ambient temperature conditions.

BS 476-32:1989   Fire tests on building materials and structures. Guide to full-scale fire tests within buildings.

BS 476-33:1993, ISO 9705:1993   Fire tests on building materials and structures. Full-scale room test for surface products.

BS ISO TR 5658-1:1997   Reaction to fire tests. Spread of flame. Guidance on flame spread.

BS ISO 5658-2:1996   Reaction to fire tests. Spread of flame. Lateral spread on building products in vertical configuration.

BS ISO 5658-4:2001   Reaction to fire tests. Spread of flame. Intermediate-scale test of vertical spread of flame with vertically oriented specimen.

BS ISO TR 11925-1:1999   Reaction to fire tests. Ignitability of building products subjected to direct impingement of flame. Guidance on ignitability.

BS EN ISO 11925-2:2002   Reaction to fire tests. Ignitability of building products subjected to direct impingement of flame. Single-flame source test.

BS ISO 11925-3:1997   Reaction to fire tests. Ignitability of building products subjected to direct impingement of flame. Multi-source test.

BS ISO 11925-2:1997   Reaction to fire tests. Ignitability of building products subjected to direct impingement of flame. Single flame source test.

BS EN ISO 9239-1:2002   Reaction to fire tests. Horizontal surface spread of flame on floor-covering systems. Determination of the burning behaviour using a radiant heat source.

In addition, there are specified fire tests for many other products and industries. The fire investigator will need to determine the appropriate tests for the particular product with which there is concern.

Appendix B: About FRS

FRS is one of the largest fire laboratories in the world and carries out fire investigations for Government and other agencies to learn lessons to improve regulations. However, FRS is only rarely involved directly with forensic investigations and so can provide independent specialist help to other investigators. FRS (which now includes the Loss Prevention Council, LPC) is the fire division of the Building Research Establishment Ltd, at Garston, Watford, UK. FRS has over 100 research and technical staff and has specialised small- and medium-scale test laboratories, Standard fire testing facilities and a 10 MW Burn Hall. Outdoor facilities are used for very large or smoky tests. FRS offers consultancy advice on all aspects of fire science that calls upon the expertise and experience of our scientific staff, and specialised research and testing facilities and fire investigation is fully integrated into the research programme. Specialists work in all the areas of fire science from materials behaviour to human behaviour and management and the work is applied to buildings, structures (offshore) and transport, including aerospace, marine, rail, road and transport infrastructure, tunnels.

The UK Office of the Deputy Prime Minister (ODPM) previously the Department of Transport, Local Government and the Regions (DTLR); previously Department of Environment, Transport and the Regions (DETR); previously Department of Environment (DoE) has long recognised the need for information from real fires in order to directly inform ministers of high profile incidents, to ensure that the guidance ODPM publishes reflects what happens during fires in real buildings, and to ensure that the whole ODPM fire research programme is supported and underpinned by information from actual relevant incidents. ODPM has funded work on both investigations into major incidents and the routine investigation of real fires by experienced staff at the FRS/BRE since the early 1970s. Consequently, the unique FRS expertise in this non-forensic type of fire investigation, developed for and supported by ODPM, has been called upon for many major inquires, including Piper Alpha (1988), the Channel Tunnel fire (1996), the Ladbrook Grove rail incident (1999) and the Yarl's Wood Detention Centre fire (2002). With the move to the functional rather than prescriptive regulations in 1985, information from real events became of even greater value to ODPM. Since 1989 this need has primarily been met by the work of the Fire Investigation Team at FRS. This team has the remit of examining fires with implications for current regulations, codes and standards, fires which have implications for current research (which itself is aimed at the development of guidance) and fires where there is a special interest by ministers or other officials. The team has access to all the research staff at FRS and, more widely, across the Building Research Establishment (BRE) and these staff can provide specialist input to an investigation about almost any aspect of the fire, including, for example, the performance of building types, ventilation systems, materials and fire-spread. The information gathered can be presented to meet particular ODPM requirements but with the primary function of giving an early warning of topics that need to be addressed during updating of documentation. This continuous review also is able to underpin the usefulness and effectiveness of current guidance by highlighting its successes.

It is this breadth of approach to understanding the implications from fires in buildings that has been of proven use. For example, investigations of factory fires contributed to identifying the problems associated with large single-storey buildings and the involvement of sandwich panel constructions, and was the basis for changes in the edition of the AD B issued in 2000. Investigations into fires involving truss-rafter roofs led to an ODPM research programme on this topic. In many other cases the effectiveness of current guidance has been demonstrated. BRE has also been able to provide similar information to the Scottish Office, the Northern Ireland Office, Home Office, DTI where appropriate.

As well as site visits by experienced FRS staff, information is gathered from a variety of sources including the police, fire brigades, forensic scientists and fire consultants in the private sector. Investigators from these other agencies are met by FRS staff both on and off site. However, all of these other bodies are almost exclusively concerned with establishing the cause, identifying the blame for the occurrence of a fire or determining liability. The involvement of FRS investigators on behalf of ODPM does not compete with any of these investigators and they have always responded positively to requests for information in the national interest of life safety. However, information is occasionally delayed if there is a court case impending and matters are *sub judice*. In order to maintain these important working relationships, FRS in general avoids commissions for private forensic investigations, but is involved in a number of UK and European fire investigation fora. Information is also obtained from literature reviews and examination of the ODPM fire statistics database.

Few, if any, other countries have adopted this continuous and systematic approach to examining the implications from generic examples of fire (see Appendix C also).

Other support for fire investigation includes demonstrations for training seminars and workshops for information transfer, the analysis of statistics (using the UK fire reporting database), computer modelling (using zone models, computational fluid dynamics, and virtual reality) and chemical and elemental analysis (including electron microscopy, ion chromatography, GC and GS–MS).

Appendix C: Fire test laboratories

The following organisations offer laboratory services that can be of assistance to fire investigators. (*Note*: forensic laboratories and overseas laboratories are not included.)

FRS (Incorporating the Loss Prevention
  Council (LPC))
Building Research Establishment
Garston, Watford, WD25 9XX, UK
Phone: +44 (0) 1923 664960
Fax: +44 (0) 1923 664910
E-mail: shippm@bre.co.uk

Health and Safety Laboratory
Process Hazards Group
Harpur Hill, Buxton
Derbyshire SK17 9JN, UK
Phone: +44 (0) 1142 892007
Fax: +44 (0) 1142 892010

Fire Safety Engineering Research and
  Technology Centre (FireSERT)
University of Ulster
Jordanstown, Newtownabbey
BT37 0QB, Northern Ireland
Phone: +44 (0) 1232 368701
Fax: +44 (0) 1232 368700

Fire Safety Engineering Group
Crew Building
The King's Buildings
University of Edinburgh
Edinburgh EH9 3JN
Scotland, UK
Phone: +44 (0) 131-650-7161
Fax: +44 (0) 131-667-9238

Fire Safety Engineering Group
School of Computing and Mathematical
  Sciences
University of Greenwich 30 Park Row
London SE10 9LS, UK
Phone: +44 (0) 208-331-8730
Fax: +44 (0) 208-331-8925

Stanger Science and Environment
Acrewood Way
St Albans,
Herts
AL4 0JY, UK
Phone: +44 (0) 1727 840580
Fax: +44 (0) 1727 816700

Warrington Fire Research Centre
Holmesfield Road, Warrington
WA1 2DS, UK
Phone: +44 (0) 1925 655116
Fax: +44 (0) 1925 655419

TTL Chiltern (TRADA)
Chiltern House
Stocking Lane
Hughenden Valley, High Wycombe
Buckinghamshire HP14 ND, UK
Phone: +44 (0) 1494 563091
Fax: +44 (0) 1494 565487

5

Modern laboratory techniques involved in the analysis of fire debris samples

Reta Newman

Introduction

When a fire investigator suspects that the cause of a fire is incendiary and that accelerants may have been involved, the forensic laboratory often becomes a valuable tool to the investigation. Samples collected in the course of an investigation and subsequently analyzed by the laboratory can provide important information to the investigator regarding the presence and/or absence of ignitable liquids or ignitable liquid residues and the chemical nature of such. Ignitable liquid detection can be used to: support the hypothesis of accelerated fire; provide investigative information regarding potential sources or uses of ignitable liquid; reinforce linkages between the fire scene and other ignitable liquids found in conjunction with the investigation; and reinforce linkages between fire scene and potential suspects.

When fire indicators at the scene indicate a poured incendiary fire, the identification of ignitable liquids residues in the associated debris can help support the investigators determination of arson. This is especially true when similar ignitable liquids are identified in multiple samples.

A comparison of an ignitable liquid residue identified in scene debris compared to an ignitable product container found in another location, can provide exclusionary information and to a certain extent relationary information. Unfortunately, the nature of ignitable liquids does not generally allow for definitive association between ignitable liquids as their analysis is generally considered class specific rather individualistic evidence. Ignitable liquids that are detected are classified based on composition. The laboratory can provide information regarding function and use of any ignitable liquids detected that may be helpful to the investigator in determining potential sources. This information is not limited to determining potential sources of accelerants but can also be useful in determining incidental sources of ignitable liquids, which can be equally important in the investigation.

Finally, investigators often isolate potential suspects at the fire scene. A suspect's apparel or possessions can be analyzed and compared to scene samples for the presence or absence of similar ignitable liquids, thus further supporting the investigation.

Laboratory analysis of fire scene samples for the presence of ignitable liquids is a four-step process consisting of sample assessment, sample preparation or extraction, instrumental analysis, and data interpretation. The instrumental analysis and resultant data interpretation are described fully in Chapter 6. This chapter will focus on the sample preparation techniques required to get the sample into a form amenable to instrumental analysis. Specifically, this chapter provides information of various sample preparation/ignitable liquid extraction techniques to include theory, application, benefits, and limitations. A discussion regarding the decision process for determining when which technique is the most appropriate will also be provided.

Sample preparation

The analysis of fire debris and related samples for the presence of ignitable liquids are accomplished with the use of gas chromatographic techniques (Chapter 6). This type of chemical instrumentation requires that the sample introduced be in a volatile liquid or vapor phase. Most samples, thus, require preparation prior to instrumental analysis to separate the components of ignitable liquids from that of the sample matrix. There are a variety of extraction/separation techniques that can be used to accomplish this purpose and no single one is necessarily ideal for all samples or situations. The most common techniques are: simple headspace sampling, adsorption, and solvent extraction.

In order to determine which technique is most appropriate to separate an accurate representation of an ignitable liquid from a given matrix, one must understand the chemical nature of the analytes. In most cases ignitable liquids can be divided in three broad categories: petroleum-derived like gasoline (petrol) and kerosene; naturally derived like turpentine and limonene; and oxygenates like acetone and alcohols. Within these categories, ignitable liquids can be further subdivided based on specific chemical composition and boiling range. A detailed description of the ignitable liquids is provided in Chapter 6, a brief overview is provided here.

Ignitable liquids

Accelerants are most often ignitable liquids that are readily available to the average consumer. Gasoline (petrol) is the most common accelerant. Other fuels, including lighter fluids, home heating fuels and diesel fuels are not uncommon. Most, including gasoline, paint thinners and solvents, and heating and engine fuels are petroleum-based products. These products are generally made up of numerous, often hundreds of compounds with individual boiling points in the range of pentane (C5) to eicosane (C20). (Note: Because most ignitable liquids are comprised of hydrocarbons, their volatility is generally described based on the elution order and boiling points of normal alkanes. The acronym "C#" refers to the normal alkane with corresponding number of carbons, for example, C8 refers to octane (C_8H_{18}), C9 refers to nonane (C_9C_{20}), etc.)

The petroleum classes of ignitable liquid can be further subdivided by more specific boiling ranges, while the majority fall somewhere within C5–C20 range, most have more narrow volatility ranges. Light products generally fall in the range of C4–C9, medium products (C8–C13), and heavy products C9–C20+ [1]. Boiling point range is a significant factor in determining appropriate sampling techniques and in optimizing individual techniques for the best recovery.

Another readily available source of ignitable liquids is that of the oxygenated solvents. These products may be found as single components or in mixtures with other chemicals including other oxygenated compounds or petroleum-based liquids. The most readily available oxygenated solvents include acetone (nail polish removers, paint cleaners, cleaning solvents), isopropanol (rubbing alcohol), and methylethylketone (paint removers, cleaning solvents). Most of the oxygenated compounds of interest to fire debris analysts are extremely volatile, more so than most of their petroleum counterparts, with boiling points at temperatures well below hexane (C6).

Recovery and detection of most oxygenated solvents, when used as accelerants, is extremely limited due to their high volatility (rapid evaporation) and water miscibility. More often than not they will be either consumed through combustion or evaporation in the course of a fire or washed from the scene in the course of fire suppression.

Oxygenated products can also be formed as incomplete combustion products so recovery of small amounts in debris samples should be considered suspect [2]. Many laboratories do not routinely screen for the presence of highly volatile oxygenated products for these reasons.

Naturally derived ignitable liquids – generally terpenes – while of different origin and having somewhat different chemical properties, are comprised of hydrocarbons and typically have boiling points in the ranges of common light to medium range petroleum products and thus are easily detected using similar sample preparation techniques. As previously noted, no single sample preparation technique is sufficient for all types of ignitable liquids in all matrices, although some techniques are more wide reaching than others and may be suitable for the vast majority of samples.

Each technique has advantages and disadvantages for various sample matrices. When selecting the best technique for a given sample the analyst must consider the sample composition, suspected ignitable liquid characteristics, the specific purpose of the analysis, the preservation of the evidence for other types of forensic analysis (i.e. fingerprints), and ignitable liquid concentration. Obviously, some of these factors cannot be determined prior to initial analysis or preliminary screening. In some instances, more than one sample preparation technique may be required to thoroughly evaluate a given sample. In these instances, the analyst must prioritize the techniques based on the destructive nature of any given procedure and the potential for best evidence. While sample assessment obviously must be the first action of the analyst in devising an appropriate protocol, at this point it is more prudent to provide a description of each of the sample preparation techniques and reserve a discussion as to which is best suited to a specific sample later in the chapter.

Sample preparation techniques

Headspace techniques

The most popular sample preparation techniques are headspace sampling techniques. There are two general types of headspace technique: simple headspace and adsorption. Headspace sampling in general, regardless of the media, provides for the cleanest samples with the least contribution from the sample matrix.

Headspace sampling takes advantage of simple evaporation and condensation principles. Ignitable liquids, by their nature, are comprised of volatile compounds. Separation from sample matrices can be achieved by vaporization of the volatile compounds from the non-volatile constituents of the sample matrix. The overall process is simple. The raw sample, commonly including solid debris, is contained in a vessel – typically a steel can, glass jar, or airtight polymer bag. Volatile compounds vaporize into the container's airspace (headspace), often aided by heat. A representation of the volatilized compounds are collected and analyzed.

If ignitable liquids are present, and the extraction parameters are properly optimized, components of any ignitable liquids present will be represented in the headspace and thus sampled. Volatilized components from the sample matrix including pyrolysis and combustion products will also be included in the headspace and will result in interfering peaks in the resultant data (Chapter 6).

The headspace can be sampled passively, where the volatiles are sampled in the sealed headspace; or dynamically, where the headspace is sampled exterior to the container facilitated by airflow. Simple headspace sampling and passive adsorption are closed system (passive) techniques. The separation of the volatile compounds and the collection of the headspace sample take place in a closed airtight vessel. Dynamic adsorption is an open system technique where external force, in the form of vapor flow, is used to facilitate the collection of the headspace sample exterior to its environment.

A closed system at a constant temperature will reach a state of equilibrium where the number of molecules entering the vapor phase (evaporating) equilibrates with the number of samples returning to the liquid phase (condensing). Evaporation/vaporization occurs when the molecules of a liquid achieve sufficient energy to escape the intermolecular forces of liquid phase into the vapor phase. Condensation occurs when the energy level of the molecule is diminished. Increases in temperature will result in increases in kinetic energy that, in turn, result in increases in the rate of evaporation for a given compound.

In a closed system, vaporized molecules will be contained in the headspace. In an open system, the molecules would be allowed to escape into an external environment. The composition of the headspace in a closed system is thus dependent on the concentration of each volatile component in the system (the mole fraction), the volatility of each component, and the temperature of the system. In fire debris analysis, the system temperature must be optimized to allow for the best possible headspace representation of volatile compounds present. Unfortunately, since ignitable liquids can contain numerous compounds with wide differences between the least and most volatile, temperature optimization can be challenging.

For a simple example consider a theoretical ignitable liquid made up of four components. The volatility range (boiling points) of the four components is wide. At a given temperature, the composition of the headspace will become constant. The rate of evaporation of each compound will equalize with the rate of condensation. At lower temperatures, compound A, with the higher vapor pressure, will be best represented in the headspace, compound D with the lower vapor pressure will be better represented in the liquid phase. The resultant analysis will show data skewed to the lighter compounds. As the temperature is increased, the energy of the molecules increase, evaporation rates for each component increase, and thus the equilibrium of each compound will shift to the vapor phase. Thus, higher temperatures favor better representation of wide boiling range mixtures.

Unfortunately, the temperatures required to fully volatilize the vast boiling range of the wide spectrum of ignitable liquid products are not practical. Higher temperatures can induce increased matrix decomposition resulting in additional interfering pyrolysis compounds in the sample headspace. Higher temperatures can also cause elevated vapor pressure (especially in the presence of water) that can result in violent container failures, which compromise the integrity of the sample and pose serious safety risks to the analyst. Higher temperatures will also induce greater displacement or breakthrough in adsorption procedures resulting in skewed data. This will be further described later. The analyst must optimize the temperature for the best representation of the headspace with an understanding that it may not be a true representation of the raw ignitable liquids present.

Every sample preparation technique has limitations. Some techniques may favor lower boiling compounds where others favor higher boiling point compounds. Each technique has specific parameters that must be considered, and in some instances adjusted, for the composition of a given sample. These limitations should not diminish the analyst's ability to accurately identify and classify ignitable liquid residues as long as the analyst is cognizant of the limitations of each sampling technique and interpret the resultant instrumental data accordingly.

Simple headspace extraction

As the name suggests, simple headspace sampling is the most basic of the headspace techniques. It is easiest and least evasive of all the sample preparation techniques. Simple headspace sampling is conducted by drawing an aliquot of the headspace into a syringe for direct injection into the gas chromatograph (Figure 5.1). The sample container is punctured and then re-sealed using tape or a septum. The sample container and the sampling syringe are brought to a constant temperature, typically in the range of ambient to 90°C (the temperature will vary with the type and concentration of the ignitable liquids in question). The syringe is heated with the sample to prevent condensation of the headspace sample in the syringe upon sampling. The syringe needle is inserted into the sample headspace and an aliquot is drawn into the syringe, which is directly injected into the gas chromatograph.

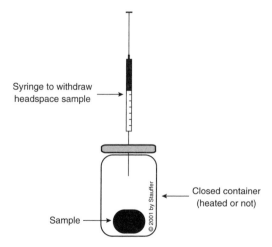

Figure 5.1 Simple headspace sampling diagram (reproduced with permission from E. Stauffer)

The composition of the analyzed vapor is dependent on the sample composition, concentration, and system temperature. Simple headspace lacks the sensitivity of the concentrated techniques and thus its efficacy as a primary sampling technique is extremely limited. Simple headspace sampling is effective for the analysis of extremely volatile ignitable liquids including low boiling oxygenated compounds like ethanol and acetone and higher concentrations of low to medium boiling (C5–C13) petroleum products. Higher boiling compounds are not typically represented in significant concentrations in the extracted sample for this technique to be effective.

The advantages of simple headspace sampling are speed and efficiency in extracting very low boiling point compounds. The disadvantages include lack of sensitivity, especially compared to adsorption techniques, and inefficiency in sampling higher boiling point compounds.

Adsorption theory

Introduction

The most widely used sample separation techniques are adsorption processes. Adsorption is the concentration of gas or liquid molecules on the surface of a solid – appropriately called an adsorbent. In fire debris analysis adsorption is almost exclusively a headspace sampling process, although analytes can be separated directly from aqueous liquid samples as well. In a headspace application, an adsorbent is introduced to the headspace of a sample container and a representation of any volatile compounds present is collected and concentrated. The adsorbent is then removed and the adsorbed species are desorbed and analyzed.

Adsorbents have chemical properties that make them amenable to trapping and retaining different types of chemical compounds. Since most ignitable liquids are comprised of mixtures of nonpolar hydrocarbons and a great many of the sample matrices are wet due to fire suppression techniques, a hydrophobic, nonpolar adsorbent is ideal. The most commonly used adsorbents for ignitable liquid extractions are activated charcoal and Tenax. Other adsorbents, including various solid phase micro extraction (SPME) fibers, enjoy limited use.

The actual composition of the final adsorbed and analyzed species is dependent on a number of factors including adsorbent properties, adsorbate properties, and system properties. The adsorbent properties are chemical composition, surface structure, and volume. The adsorbate properties are chemical structure to include functional groups as well as size, and volatility. The system properties include headspace volume, matrix composition, and most importantly, temperature.

The adsorbents attach the adsorbate by weak intermolecular forces. This is a physical adsorption process where the chemical structure of the adsorbed species is unchanged. Absorption is an instantaneous but not irreversible process. When an adsorbate comes into contact with the adsorbent it will attach (become adsorbed), how long it will stay attached is dependent on its adsorptivity, which is a function of the aforementioned factors and which are described next.

Adsorbent properties

First and foremost, the selected adsorbent must favor nonpolar compounds. The activated charcoal, also called active carbon, adsorbent used in fire debris analysis is generally produced by the destructive distillation of coconut shells. The resultant surface structure consists of large and small nonuniform fissures called macro and micropores. The larger, macropores, act as conduits to the micropores where adsorption occurs. Active carbon is especially suited for adsorption of nonpolar hydrocarbons. Active carbon is suitable for the collection compounds in a very wide boiling range (~0–260°C) [3]. The efficacy of the carbon adsorption decreases with increasing polarity, thus more polar compounds (generally produced from the sample matrix) present in the headspace are less efficiently adsorbed.

Tenax TA is a synthetically produced porous polymer based on the 2,6-diphenylene oxide compound [4]. While the effective recovery using Tenax is more limited than that of activated charcoal (~50–300°C) [5], it is well suited for the compounds of interest to the fire debris analyst as the boiling range of C5–C20 hydrocarbons lie between 35°C and 300°C. In comparison, Tenax TA is slightly more efficient in the recovery of higher molecular weight compounds while activated charcoal has better lower end recovery. As a result, the extraction of identical samples, with all other factors being equal, Tenax TA may result in data favoring the higher boilers than that of charcoal, however, these differences are generally accounted for in the overall analysis scheme and resultant data interpretation and should not affect the analysis process. Both Tenax and activated charcoal have been shown through validation and application to be appropriate and effective adsorbents for most classes of ignitable liquids.

In SPME, a silica fiber is coated with the adsorbent. Like Tenax TA, the adsorbents on SPME fibers are also polymer based. Polydimethylsiloxane (PDMS) is the most commonly recommended fiber for extraction of hydrocarbons from fire debris samples [5,6]. This is the same film coating found in the nonpolar capillary columns commonly used to analyze ignitable liquids in gas chromatographs (Chapter 6). The most effective analyte boiling range for PDMS could not be found in the literature, however, given the presentation of data collected in similar trials, it appears to be higher than that of Tenax.

The second property of the adsorbent that must be considered when optimizing a procedure is the pore size or pore distribution, which is a function of the surface properties. These are the surface areas where adsorption occurs. This must be optimized to include the variety hydrocarbon configurations that boil in the range of C5–C20. Adsorbents with too small pore sizes will result in an exclusion of sterically hindered larger molecules. Adsorbents with very large pore sizes will result in increased displacement rates favoring the adsorption/retention of higher molecular weight hydrocarbons. In reality this is only a concern in selecting activated charcoal, as pore size is a uniform feature of the polymers Tenax TA and PDMS.

Finally, the amount of adsorbent present must be sufficient to collect a fair representation of the headspace. Adsorbents have limited capacities. When those capacities are exceeded the nature of the adsorbed species generally shifts to the compounds with the higher affinities for the adsorbent. In the case of active charcoal, Tenax TA, and PDMS SPME fibers, higher molecular weight compounds are better adsorbed than their lower molecular counterparts. When the capacity of the adsorbent is exceeded, a disproportionate representation of the headspace can result; this is the phenomenon of displacement. Except in extreme cases, it does not typically preclude proper identification or classification of ignitable liquid present.

Of additional concern, with activated charcoal and some SPME fibers, aromatic compounds have greater affinities than aliphatic compounds. Typically, unless the adsorbent is grossly saturated it is unlikely that this particular preference will affect recovery to an extent to prohibit proper identification of any ignitable liquids present as long as the analyst is aware of and compensates for the phenomenon in the data interpretation phase of the analysis.

Adsorbate properties

Activated charcoal is especially suitable for sampling C5–C18 hydrocarbons and performs adequately for C18–C20+ compounds when they are adequately represented in the sample headspace. While activated charcoal has been used to collect oxygenated solvents [7], both their the lower boiling point and the increased polarity of these compounds would indicate that other adsorbents or extraction techniques would be better suited for more accurate and proportional representation.

Tenax TA is suitable for both polar and nonpolar analytes, however, it is less efficient in the lower boiling range where many of the common oxygenates of interest are found. The same situation is true for PDMS SPME fibers, however, alternative more polar fibers including the Carboxen/PDMS fibers, are available and appear to work well [8].

Once an adsorbent is selected, the headspace composition, concentration, and system temperature will be the biggest determinants for the final composition of the adsorbed species [9]. Multiple extractions with differing parameters may be necessary in some situation to obtain the best representative data.

System properties

The headspace volume must be such that there is sufficient room for the volatile compounds to diffuse. Insufficient headspace generally favors lower boiling point compounds, which can result in a significant shift in apparent boiling range of the ignitable liquid [10]. This is a notable problem when nonrigid containers (polymer evidence bags) are used for sample collection. If the headspace of the container is not sufficient, the lower boiling compounds can swamp the headspace. Inflating the headspace with air is recommended to ensure adequate headspace volumes.

The matrix composition is another significant factor in the adsorption of ignitable liquids from fire debris. Absorbent and adsorbent materials in the sample matrix will inhibit volatilization. Charred debris, while not nearly as efficient, has the similar adsorptive properties as activated charcoal. In general, charred debris will retain higher molecular weight compounds and, in some cases, aromatic compounds and thus these types of compounds may be less represented in the headspace and will not be proportionately represented in the analyzed sample. In general, in the presence of charred debris, the adsorption of compounds that elute above C18 are not consistently or proportionately represented. There are, of course, exceptions when the ignitable liquid concentration is very high or the sorption properties of the matrix are not as great, but as a general rule, an analyst cannot differentiate a kerosene range heavy petroleum product (C9–C18) from a diesel range heavy petroleum product (C9–C25) using adsorption techniques alone [10].

And, finally, perhaps the most significant system factor is that of temperature. The adsorption temperature impacts the composition of the final adsorbed species in two competing ways: higher temperatures favor volatilization of the best representation of compounds in wider boiling ranges of components common to petroleum-based ignitable liquids; lower temperature slow the rate of displacement that can distort the relative concentration of individual components in the adsorbed species. Elevated temperatures will result in the preferential desorption of lower boiling compounds thus shifting the composition of the final adsorbed species to the higher boiling compounds. Adsorption headspace techniques generally use temperatures in the range of 60–80°C as a compromise to these two effects.

Adsorption temperatures must be such that a reasonable representation of ignitable liquid components in the boiling range between C6 and C20 is achieved to allow for trapping on the adsorbent. Ambient temperatures are generally sufficient for moderate to high concentration of light to medium petroleum products, however, additional heat is typically needed to sufficiently volatilize components of medium to heavy products. While higher temperatures result in greater headspace presence overall and better representation of less volatile compounds in general, they do not prove beneficial and in fact can be detrimental to adsorption extractions.

As previously stated, some of the concerns of higher temperatures are associated with dangerously high vapor pressures and the increased potential for thermal degradation of the matrix. The adsorption process itself is also impacted by heat. The additional determinant of elevated temperatures is increased preferential adsorption. And again, while this is not typically so significant as to preclude proper conclusion when moderate temperatures are used, the displacement rate increases significantly at higher temperatures and can preclude proper sampling and resultant ignitable liquid analysis data.

Increased temperatures result in higher kinetic energy in the molecules. Additional energies result in shorter residence times overall and accelerates the displacement process as more volatile molecules will thermally desorb faster and become replaced by dispropor-tionate numbers of less volatile molecules. Thus, the adsorption sampling temperature must be balanced to allow for the best possible representation of volatiles in the headspace from the initial sample and the best representation of the headspace in the final adsorbed sample. Generally temperatures in the range of 60–80°C are considered ideal for general screening and sampling of the various ignitable liquid classes. Exceptionally strong highly volatile samples (i.e. lighter petroleum products, gasoline, etc.) can often be accurately sampled at ambient temperatures.

Adsorption techniques

Passive activated charcoal adsorption

The most common extraction technique used in the United States is, by far, adsorption on activated charcoal. Activated charcoal made from coconuts has been found to be the most effective for extraction of C6–C18 hydrocarbons. Activated charcoal can be used in either passive or dynamic system configurations. In a passive system, headspace vapors contact the adsorbent by simple diffusion. In a dynamic system vapor flow is introduced into the system to force the contact between the adsorbent and adsorbate.

Passive headspace concentration is a very simple process with equally simple apparatus. Activated charcoal, generally in a form impregnated on an inert polymer strip is suspended in the headspace of the sample container (Figure 5.2) [11]. Molecular transport is by simple diffusion, volatilized molecules randomly bounce around in the headspace until they collide with the adsorbent and become adsorbed. The composition of the headspace is a function of the volatility of the individual components and system temperature as described earlier.

A molecule will stay adsorbed until it achieves sufficient energy to escape back into the vapor headspace. This is analogous to molecules of a liquid escaping to the gas phase when

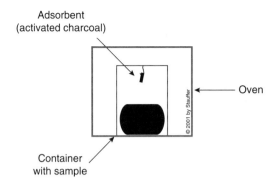

Adsorbent
(activated charcoal)

Oven

© 2001 by Stauffer

Container
with sample

Figure 5.2 Passive headspace adsorption with activate charcoal impregnated on a polymer strip (reproduced with permission from E. Stauffer)

sufficient energy is achieved. The difference, in terms of hydrocarbon adsorption on charcoal is that adsorption forces are only approximately 2–3 times that of condensation forces. Thus, as higher molecular weight hydrocarbons are less volatile than their lower molecular counterparts, the higher boiling compounds have a greater affinity for the adsorbent.

Adsorbents have a finite capacity that is related to the surface structure and gross volume of adsorbent. If the concentration of vapors in the headspace is such that the adsorbent capacity is exceeded, a displacement of heavier versus lighter compounds can occur. Less strong, but potentially significant is the affinity of aromatic compared to aliphatic compounds. In excessively strong samples, the potential exists for aromatic compounds to be preferentially represented in the final adsorbed species. Except in very extreme samples, displacement does not affect the overall data interpretation as long as the analyst is aware of and compensates for displacement effects in data comparison and interpretation. The parameters associated with passive headspace concentration using activated charcoal and that can directly impact the success of the adsorption in extracting a truly representative sample of the sample headspace are: temperature, duration, adsorbent properties, volatile component concentration, and desorption solvent [11].

Once the adsorption temperature is established, typically in the range of 60–80°C, the duration must be addressed. This must be balanced with the temperature. Adsorption time is defined as the amount of time that the activated charcoal remains in contact with the headspace. Passive headspace concentration is very forgiving in that broad ranges in sampling time typically result in minor differences. The duration is generally a function of adsorption temperature, volatile composition, and concentration. While volatile composition and concentrations are generally considered unknown, strong samples of lighter products (i.e. gasoline) have obvious distinctive odors. Often these sample can be extracted at a lower (ambient) temperatures or with shorter durations (2–4 h at standard temperatures) to minimize displacement effects. Typical durations for 60–80°C routine sampling are commonly in the range of 8–18 h [11]. It is common practice for an analyst to place the adsorbent in the sampling container and place the container in the oven at the end of the workday and remove and extract it the following morning.

Once the analyte is adsorbed, it must be removed (desorbed) for instrumental analysis. There are two ways in which adsorbate can be removed from adsorbents: thermally or chemically. Thermal desorption is not practical for activated charcoal as the extremely high temperatures required are not practical. Thus activated charcoal is chemically desorbed.

In order to be effective, the desorption solvent must be such that the adsorbed species is readily soluble and it must have a high adsorptivity to bind the active site of the adsorbent and prevent selective retention of stronger adsorbing compounds in the volatile mixture. Carbon disulfide has been shown to be, by far, the best desorption solvent resulting in the best representation of the adsorbed species [12]. Unfortunately, CS_2 is a very dangerous toxic chemical. Alternatives, including diethylether, pentane, and dicloromethane are also used [12–14]. All result in lesser recoveries and, to some extent, preferential desorption of aliphatic over aromatic compounds. In most situations, these variations are insignificant and analysis and data interpretation are not significantly affected when interpreted by a well-trained analyst.

The advantages of passive headspace sampling with activated charcoal are many. First, the technique is essentially nondestructive, only a small amount of the total headspace is removed, allowing for additional techniques or repeat analysis as needed. The exception would be simple non-sorbent matrices (like glass) with very small amounts of ignitable liquid present. In that situation it is possible to effectively volatilize and remove all of the ignitable liquid in the initial extraction [15].

Another advantage is the ease of sample preservation. A portion of the adsorbed charcoal can be retained for future (defense, insurance, quality control) testing. While ignitable liquid evidence containers are generally considered airtight, they often are not ideal for long-term storage; steel cans rust, polymer bags puncture, and glass jars can break. Overtime, and in the process of accessing the evidence, the airtight integrity of the containers is often compromised. Extract preservation allows for future analysis and alternative for evidence preservation [16].

Additionally, passive headspace adsorption on activated charcoal is very efficient in terms of analyst participation. The amount of time required to suspend the adsorbent, place the container in an oven, remove it, elute it with a solvent is generally less than 5 min. It is a very forgiving technique; where the difference between 12 and 18 h adsorption times is usually negligible. Thus, periodic inattention by the analyst will not result in sample destruction nor will it affect recovery, and of course, reanalysis is always an available option.

Since all of the equipment used to prepare and store the extract is disposable, quality is very easy to control and monitor. Since no complex equipment is required, numerous items of evidence can be extracted simultaneously.

There are disadvantages to this technique as well. Solvents used for elution are often toxic or dangerous. The use of solvents for desorption serves to reduce the sensitivity of the method as only a dilute portion of the recovered volatiles is actually injected into the gas chromatograph and analyzed. Thermally desorbed adsorbents like Tenax and SPME fibers are thus more sensitive.

The length of time involved in the process can be a negative factor as well. While this procedure requires very little analyst time or activity, the overall process takes more time per individual sample than other separation techniques.

Dynamic activated charcoal adsorption

Until activated charcoal strips were widely available, the most common way to use activated charcoal for ignitable liquid recovery was in a dynamic system. In a dynamic system the headspace is forced to the adsorbent, exterior to the sample container, with air or nitrogen flow or a vacuum. In the most common configuration, loose activated charcoal is placed in two glass Pasteur pipettes using glass wool plugs. Two holes are punctured in the sampling container. The pipettes are secured in the holes with septa. The container is heated (70–90°C) to volatilize any ignitable liquid components and a vacuum is drawn through, or air or nitrogen is flown through one of the pipettes flushing the headspace into contact with the activated charcoal (Figure 5.3). The second pipette serves to filter the added air that is drawn or pushed through the container. The adsorbed charcoal is removed from the pipette and eluted with a solvent. Again, like passive headspace concentration, carbon disulfide is most commonly used and generally recommended, however, pentane, diethylether and dichloromethane are also used.

The adsorption dynamics are somewhat similar to that of passive adsorption with one significant difference. In passive adsorption molecular transport is by simple diffusion, in dynamic molecular transport it is by diffusion and flow. The result is that the overall sampling process is much faster and the rate of adsorption/desorption is much faster. The addition of flow introduces a new phenomenon – breakthrough. Breakthrough is the loss of lighter, more volatile components from the adsorbent. It is somewhat similar to the displacement that occurs in passive systems. In a passive system the rate of displacement is a factor of temperature. In a dynamic system the rate of displacement is affected by both temperature and flow rate. The additional kinetic energy provided by the flow will greatly increase the adsorption/desorption rate and thus the flow rate must be carefully controlled. Also, in a passive system, when a compound is displaced (desorbed and replaced with a stronger attaching molecule), it is returned to the headspace. In the case of breakthrough the displaced samples are drawn through and out of the system – never to be sampled again. As a result, dynamic headspace sampling with activated charcoal is much less forgiving than passive sampling. It can be a destructive technique and is much less amenable to repeat analysis.

The advantages of dynamic sampling with activated charcoal are speed and sensitivity. Passive sampling typically requires 8–18 h adsorption times and while there is very little analyst involvement, it precludes quick results. Dynamic sampling can be done in 20–30 min, it requires the complete attention of the analyst during that period but results for individual samples can be generated fairly quickly. Additionally, because this technique

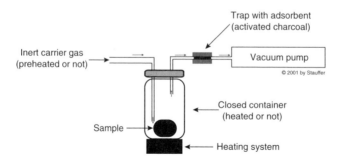

Figure 5.3 Dynamic headspace sampling with activated charcoal (reproduced with permission from E. Stauffer)

results in flushing the headspace rather than passive sampling, it is more sensitive. Like passive adsorption, dynamic activated charcoal adsorption technique allows for long-term retention of the sample extract as an alternative for sample preservation.

The disadvantages of this technique include the use of toxic desorption solvents, more rigid parameters, and its generally destructive nature. This technique requires total analyst involvement and attention from start to finish. Parameters must be highly regulated and monitored to prevent sample loss and destruction. Additionally, this technique is not typically amenable to 'batching', that is, only one item of evidence can be sampled at a time, which results in an overall decrease in efficiency and productivity.

Tenax adsorption

Another useful, and widely used adsorbent for volatile hydrocarbon concentration is Tenax TA. Tenax is widely used by fire debris analysts in Canada and Europe. The sampling apparatus for the use of Tenax is not complex. In the simplest configuration, loose Tenax is placed in a sampling tube. A large disposable syringe barrel is attached to one end of the tube; a syringe needle is attached to the other. A hole is punctured in the sample container and secured with tape or septa. The container is heated to vaporize compounds of interest (60–80°C). The Tenax syringe apparatus is inserted into the container and 30–50 mL of the headspace is drawn to the adsorbent (Figure 5.4). The metal sampling tube is removed from the container and the syringe is dissembled. The Tenax is then thermally desorbed at the injection port of the GC [17].

While the Tenax technique is technically dynamic adsorption, it has many of the advantages of passive headspace sampling. Like dynamic adsorption with activated charcoal, molecular transport is by both diffusion and flow. Unlike the dynamic activated charcoal system, only a small portion of the headspace is drawn to the adsorbent, thus this technique is less destructive and generally allows for repetitive analyses.

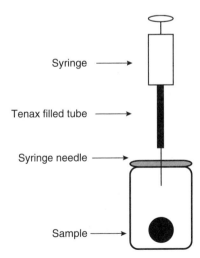

Syringe

Tenax filled tube

Syringe needle

Sample

Figure 5.4 Dynamic headspace sampling with Tenax

Sampling is much faster and requires less analyst involvement than dynamic charcoal adsorption. The sample can be extracted as soon as it reaches the equilibrium temperature and the sample process of drawing headspace into the syringe only takes a few minutes. Since the adsorbed species is thermally removed and 100% goes into the GC, the Tenax adsorption process is significantly more sensitive than activated charcoal, which requires solvent desorption. If a portion of the adsorbed Tenax is removed prior to thermal desorption, it can be retained as a long-term alternative for sample preservation.

The only significant drawbacks to Tenax sampling is the increased propensity for displacement of early eluting compounds and the requirement for specialized equipment in the form of thermal desorption apparatus for the gas chromatograph.

Solid phase micro extraction

Another type of adsorbent available to the fire debris analyst is the SPME fiber. SPME is a relatively new technique for sampling a variety of compounds from a variety of matrices [8]. SPME fibers are constructed of silica fibers coated with an adsorbent polymer. For the purposes of fire debris analysis, two fibers have gained the attention of chemists. The first is the polydimethylsiloxane (PDMS) fiber. This is a nonpolar fiber that is recommended for the extraction of nonpolar compounds like those found in petroleum products [18]. The second is the more polar Carboxen/PDMS cross-linked fiber, which has been shown to be effective for sampling lower molecular oxygenated solvents [19].

SPME sampling requires a special apparatus designed to hold, protect, and expose the SPME fiber. Once the sample container is brought to temperature, the sample holder, which somewhat resembles a syringe, is inserted into the headspace. The fiber is exposed to the headspace for a short period of time, typically less than 10 min. The fiber is then retracted into a protected needle in the holder and the holder removed from the sample container. The needle is inserted into the injection port of the GC where the fiber is exposed and thermally desorbed (Figure 5.5).

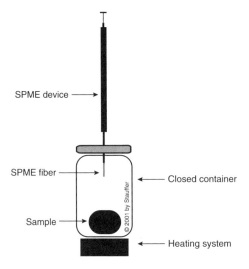

Figure 5.5 Passive headspace sampling with SPME fiber (reproduced with permission from E. Stauffer)

The biggest deterrent for the use of SPME fibers is the extremely limited sample capacity. Currently, the fiber with the greatest capacity is the 100 μm PDMS. Because this provides relatively few adsorption sites, displacement occurs very quickly. In most cases, SPME extractions will favor, often heavily, the higher molecular weight hydrocarbons and compounds in the headspace. Also, like charcoal, SPME fibers have been shown to preferentially adsorb different types of hydrocarbons (aromatics vs aliphatics), which could potentially affect data interpretation in some cases [20].

Because the sampling time is such a narrow range (typically 5–10 min) it is hard to optimize the system for ideal recovery. Fortunately, the sampling process is very easy and, because this is a passive system, it can be done repeatedly without significant alteration to the sample. Thus, if an initial extraction appears to be heavily displaced, the sampling time and/or temperatures can be optimized to the situation, and the item quickly reanalyzed with better results.

Solvent extraction

An alternative, or supplement, to the various headspace sampling techniques is that of solvent extraction. Solvent extraction was one of the first techniques used for ignitable liquid residue recovery from debris and still enjoys limited use today. The theory of solvent extraction is almost self-explanatory. The sample (debris) is washed with an appropriate solvent to remove any ignitable liquid components. Typically large volumes of solvent are required to thoroughly extract the sample and therefore the wash is concentrated prior to instrumental analysis. The most common solvents are pentane, carbon disulfide, and methylene chloride. Pentane is less toxic but is difficult to obtain in pure form. Carbon disulfide is cleaner but very toxic. Methylene chloride is a more universal solvent, which tends to result in more significant amounts of matrix contribution in the analyzed sample.

Unfortunately, not only are ignitable liquid components recovered in the solvent wash, but also dissolved compounds in the sample matrix. As a result, solvent extraction is less sensitive, less selective, and much dirtier than adsorption techniques. Additionally, the more volatile compounds are often lost in the concentration process making this technique less desirable for lighter petroleum products.

Solvent extraction does have a place in the modern analysis scheme for ignitable liquid recovery. For simple, nonporous or less porous and nonsoluble matrices, solvent extraction results in fast and good results. For complex porous and soluble matrices, the use of solvent extraction is generally limited to supplemental extractions used to differentiate kerosene range products from diesel fuel range products when headspace sampling techniques are used. Even then, its efficacy is limited due by substrate contribution and ignitable liquid concentration.

Conclusion

With so many options, the question becomes, what extraction technique should be used in which situations? Most laboratories use a form of adsorption, typically activated charcoal, or Tenax, for most analysis. Which adsorbent and in which system is determined by laboratory policy and personal preference. In general passive headspace sampling with activated charcoal or dynamic sampling with Tenax are recommended. Simple headspace sampling is often used prior to adsorption for general screening. It is also usually the technique of choice when analyzing for the presence of low molecular oxygenated solvents. SPME has not gain widespread use, however, as more interest is generated for it, this is likely to change. Currently, its use is limited to that of simple headspace sampling – extraction of low molecular oxygenates and general screening. Neither is typically used as a lone or primary technique. Solvent extraction is used for rapid analysis of simple matrices and for differentiation of ignitable liquids in the heavy petroleum product range when such differentiation is desired. Unfortunately, there is, as yet, no one technique or set of associated parameters that is ideal for extracting ignitable liquids from all samples, however, the popular adsorption techniques come close. In the end, the training and experience of the analyst, coupled with the circumstances of the fire investigation and the needs of the investigator will determine which methods are the most appropriate for each individual sample.

References

1   American Society for Testing and Materials (2002) ASTM E 1618-01 test method for identification of ignitable liquid residues in extracts from fire debris samples by gas chromatography–mass spectroscopy, in *Annual Book of ASTM Standards, 2002.*

2   B. Levin (1986) A summary of the NBS literature reviews on the chemical nature and toxicity of the pyrolysis and combustion products from seven plastics. U.S. Department of Commerce, Washington.

3   R. Augustin, H. Bittner, and H. Klingenberger (2000) Volatile organic compounds from adhesives and their contribution to indoor air problems, Gerstel Application Note, Hamburg, Germany.

4   Scientific Instrument Services (2002) Tenax TA adsorbent resin physical properties, sisweb.com

5   K. Furton, J. Almirall, and J. Bruna (2000) The use of solid phase microextraction–gas chromatography in forensic analysis *Journal of Chromatographic Science* 38: 297–306.

6   A. Harris (2003) GC-MS of ignitable liquids using solvent desorbed SPME for automated analysis. *Journal of Forensic Sciences* 48(1): 41–46.

7   J. Phelps, C. Chasteen, and M. Render (1994) Extraction of low molecular weight alcohols and acetone from debris using passive headspace concentration. *Journal of Forensic Sciences* 39(1): 194–206.

8   Supleco (1998) Solid Phase Microextraction, Supelco Product Information Guide, Sigma-Aldrich, St Louis.

9   R. Newman, W. Dietz, and K. Lothridge (1996) The use of activated charcoal strips for fire debris extraction by passive diffusion. Part I: The effects of time, temperature, strip size and concentration. *Journal of Forensic Sciences* 41(3): 361–370.

10 M.L. Fultz (1995) Analysis protocols and proficiency testing, Proceedings of the International Symposium on the Forensic Aspects of Arson Investigations, 165–194.

11 W. Dietz (1990) Improved charcoal packing recovery by passive diffusion. *Journal of Forensic Sciences* 35(2): 111–121.

12 J. Dolan and R. Newman (2001) Solvent options for the desorption of activated charcoal in fire debris analysis, presented at the American Academy of Forensic Sciences, Seattle, Washington.

13 J. Lentini and A. Armstrong (1997) Comparison of the eluting efficiency of carbon disulfide with diethyl ether: the case for laboratory safety. *Journal of Forensic Sciences* 42(2): 307–311.

14 G. Hicks, A. Pontbriand, and J. Adam (2003) Carbon disulfide vs dichloromethane for use of desorbing ignitable liquid residues from activated charcoal strips presented at the American Academy of Forensic Sciences, Chicago, Illinois.

15 R. Newman (1996) An evaluation of multiple extractions of fire debris by passive diffusion, Proceedings of the International Symposium on the Forensic Aspects of Arson Investigations, 287.

16 L. Waters and L. Palmer (1993) Multiple analysis of fire debris using passive headspace concentration. *Journal of Forensic Sciences* 38(1): 165–183.

17 D. Deharo (2003) Personal communication.

18 K.G. Furton, J.R. Almirall, and J.C. Bruna (1996) A novel method for the analysis of gasoline from fire debris using headspace solid-phase microextraction. *Journal of Forensic Sciences* 41(1): 12–22.

19 C. Woodruff and R. Newman (2002) Analyzing trace amounts of low molecular weight oxygenates from fire debris using solid phase microextraction, pending publication.

20 J. Lloyd and P. Edmiston (2003) Preferential extraction of hydrocarbons from Fire Debris samples by Solid Phase Microextraction. *Journal of Forensic Sciences* 48(1): 130–134.

6

Interpretation of laboratory data

Reta Newman

Introduction

Ignitable liquid identification in samples collected in a suspected arson can provide valuable information to fire investigators. While the determination as to the cause, origin, and incendiary nature of a fire is the responsibility of the fire investigator, not the forensic laboratory, the laboratory can provide information to the investigator to help, support, or in some cases disprove, investigative hypotheses; isolate potential sources of ignitable liquids; and help establish important investigative links among cases, scene samples, and suspects.

The laboratory techniques used for the extraction and analysis of volatile compounds can be very sensitive, allowing for the identification of minute amounts of ignitable liquids in fire debris. The identification of an ignitable liquid in a scene sample is not typically sufficient, in itself, for determining that a fire was incendiary. Conversely, the lack of identification of an ignitable liquid does not preclude that a fire was not incendiary or that an accelerant was not used in its perpetuation.

The primary role of the forensic laboratory in a fire investigation is the analysis of scene samples for the presence of ignitable liquids. An ignitable liquid is simply a flammable or combustible liquid, generally a liquid with a flash point of less than 200 F. Ignitable liquids have widespread and broad based use in a multitude of applications including engine fuels, home heating fuels, charcoal starters, polishes, insecticides, transfer vehicles, cleaning solvents, paint thinners, and lubricants which represent hundreds of thousands of commercially available products, any of which can be used as an accelerant. An accelerant is anything used to intentionally start or spread a fire. An accelerant is often an ignitable liquid; however, an ignitable liquid, even when identified in fire debris, is not necessarily an accelerant. When kerosene is used as a home heating fuel it is an ignitable liquid. When it is intentionally poured through a structure and ignited with a match it is an accelerant. From a chemical perspective they are identical.

The identification of an ignitable liquid in any given fire related sample does not, in itself, imply that a fire was incendiary or that an ignitable liquid identified was an accelerant.

Gasoline identified in a debris sample taken from a garage where a gasoline can was stored is certainly an ignitable liquid, but probably not an accelerant as it is likely to be incidental to the scene. Incidental sources for ignitable liquids must be eliminated and/or other aspects, most importantly the origin and cause of investigation at the scene must corroborate the laboratory's findings before it can be determined that any identified ignitable liquid is an accelerant.

Determining the significance of laboratory results requires thorough understanding of ignitable liquids including manufacture, composition, chemical properties, and product uses. Additionally, an extensive knowledge of sample matrices from which ignitable liquid residues are extracted, including types of incidental ignitable liquids that may be common to the matrix, and types of pyrolysis and combustion products formed when the matrix is exposed to fire is also important.

And finally, the limitations of sample extraction techniques as described in Chapter 5, must be understood and taken into consideration in both ignitable liquid identification and in sample comparison. Comparison of sample extracts to neat liquids will often result in subtle differences that can lead to false conclusions regarding potential product sources and sample relationships. All of these concepts will be provided in this chapter, because the intended audience is very broad, both basic definitions and complex applications are included.

Ignitable liquid composition

General classifications

Ignitable liquids, any of which that can be used as an accelerant, can be broadly categorized as petroleum or nonpetroleum. Petroleum-based ignitable liquids are refined from crude oil; examples include gasoline (petrol), kerosene, and diesel fuels. Petroleum-based ignitable liquids are comprised of hydrocarbons. Hydrocarbons are chemical compounds with a molecular structure limited to carbon and hydrogen. Nonpetroleum-based ignitable liquids are those derived from other sources. They include oxygenated solvents and naturally derived products (turpentine). The analysis processes for ignitable liquids of the two classes are similar; a distinction is made between them because there are distinct differences in sample chemistry resulting in different approaches to the related data interpretation.

Hydrocarbon chemistry

Petroleum-based ignitable liquids contain alkane and/or aromatic hydrocarbon compounds. Alkanes are saturated carbon and hydrogen molecules in one of three structural designations: normal alkanes, isoalkanes, and cycloalkanes. The terms "saturated" and "unsaturated" refer to the type of bonds present in a molecule. Saturated compounds exclusively contain single bonds. Unsaturated compounds contain at least one double or triple bond. Unsaturated compounds are generally more reactive than their saturated counterpart. The general structures and common examples of the various types of hydrocarbons described are provided in Table 6.1.

Normal alkanes (*n*-alkanes) consist of a series of carbon atoms connected by single bonds in a straight-chain configuration. None of the carbon atoms in a normal alkane molecule is attached to more than two other carbon atoms. *n*-Alkanes are often used as retention time or elution order markers for ignitable liquid data interpretation. For simplicity, fire debris

Table 6.1 Common examples of various types of hydrocarbons

Type of compound	Specific structure	Description	Examples
Alkane	Normal alkanes (n-alkanes)	Straight-chain hydrocarbons (C_nH_{2n+2})	
	Isoalkanes (isoparaffins)	Branched-chain hydrocarbons (C_nH_{2n+2})	
	Cycloalkanes (cycloparaffins, naphthenics)	Cyclic hydrocarbons (C_nH_{2n})	
Aromatic	Simple	Alkyl-substituted benzenes	
	Polynuclear (naphthalenes)	Multiple fused benzene ring compounds	
	Indane	Benzene ring fused to cyclopentane	

analysts generally use the acronym "C#" for a given normal alkane. For example, *n*-pentane (C_5H_{12}), a five-carbon normal alkane is often referred to as C5; *n*-tridecane, a thirteen-carbon compound, is referred to as C13.

Isoalkanes are also called isoparaffins. Like *n*-alkanes, their molecular formula is C_nH_{2n+2}, where *n* is the number of carbon atoms. They differ from *n*-alkanes in the configuration of bonded carbon atoms. The structure of an iso-alkane includes at least one carbon bonded to three or four other carbons. This branching structure results in compounds with lower boiling points than their *n*-alkane counterparts.

Normal and isoalkanes are aliphatic compounds. The term "aliphatic" refers to straight-chain and branched hydrocarbons and generally includes both saturated and unsaturated molecular configurations. For the purposes of this work, the usage of the term aliphatic is generally restricted to the saturated compounds, that is, normal and isoalkanes.

Alkenes are unsaturated hydrocarbons. They are similar to alkanes but with the presence of a double bond in the molecular configuration. Alkenes are not commonly found in ignitable liquids as they are extremely reactive. They are, however, often found in pyrolysis products of sample matrices and their presence can be an important diagnostic feature in the interpretation of data from fire debris extracts.

Cycloalkanes (cycloparaffins) are saturated hydrocarbons configured in a cyclic structure and with a chemical formula of C_nH_{2n}. Unlike isoalkanes, cycloalkanes have higher boiling points than the corresponding normal alkanes. As will be described in greater detail later, the presentation of data used in fire debris analysis is generally based on the boiling points of the individual compounds. Most of the cycloalkanes of interest to fire debris analysts are based on the cyclohexane molecule.

157

Aromatic compounds contain the benzene ring structure (a six-member ring with alternating double bonds). The structural designations of aromatic compounds that are used in ignitable liquid data interpretation are alkylbenzenes, polynuclear aromatics (PNA), and indanes.

The simplest aromatic compounds found in petroleum-based ignitable liquids consist of alkyl (alkane)-substituted benzene compounds (alkylbenzenes). Benzene itself is generally removed or minimized in most products due to health and environmental concerns. Successive alkyl substitutions on the benzene structure are generally described as "C# alkylbenzene." For example 1,2,4 trimethylbenzene and *n*-propylbenzene are benzene molecules with three-carbon substitutions, and thus are called C3 alkylbenzenes. The aromatic constituents of petroleum-based ignitable liquids, when present, are very diagnostic. The aromatic compound peak groups are based on the degree of substitution. Generally, C2–C4 alkylbenzenes are of the most interest to fire debris analysts.

PNAs are comprised of two or more benzene rings fused together. Generally, naphthalene, 2-methylnaphthalene, and 1-methylnaphthalene are the only PNAs found in any significant amounts in ignitable liquids. The presence and relative ratios of naphthalene, 2-methyl-naphthalene, and 1-methylnaphthalene can be used in ignitable liquid data interpretation and classification, however, they are also compounds which are commonly produced in the pyrolysis of synthetic materials and thus tend to be less important indicators than other classes of aromatic compounds.

Indanes are aromatic compounds with an alkyl side chain attached in a cyclic configuration to the benzene ring. The side chain attaches to the ring in two consecutive locations resulting in a cyclic configuration. Although not typically abundant, the presence of C1 and C2 substituted indanes can be used to help distinguish pyrolysis from some ignitable liquids.

Petroleum-based ignitable liquids

Petroleum-based ignitable liquids are the most common and represent the largest number of products. These products are refined from crude oil, as opposed to petrochemicals, which are individual compounds that are separated from petroleum for use in the production of other chemical products, most notably polymers. Crude oil is refined to produce products with specific characteristics; the final composition of these products is based upon the intended market. Gasoline, the most abundant refinery product, is produced to meet the specifications of the automobile industry and various regulatory agencies. Distillates are produced for use as fuels and solvents. Additionally, in an effort to make the most efficient use of refinery waste streams and to generate as many useful products from a barrel of crude oil as possible, a variety of specialty products are produced and marketed for additional uses. A detailed explanation of petroleum refinery processes used to create the various products is beyond the scope of this work, however excellent texts are available on this subject [1,2].

The fire debris analyst is typically interested in petroleum products in the boiling range of *n*-pentane (C5) to eicosane (C20), with boiling points between 36–205°C. The common classes of petroleum products in this range are gasoline, petroleum distillates, isoparaf-finic products, naphthenic–isoparaffinic products, aromatic solvents, and normal alkane products [3]. With the exception of gasoline, these classifications are based upon chemical composition rather than end-use.

Gasoline is comprised of hundreds of compounds taken from several refinery processes and blended together to meet required specifications. Designed for use in internal combustion engines, large amounts of aromatic compounds are included in gasoline to increase engine performance and fuel efficiency. The end product is a volatile liquid with very distinctive chemical properties.

Petroleum distillates are products of crude oil distilled between two temperatures. Distillates contain an abundance of aliphatic compounds, typically with predominant normal alkanes and less significant aromatic compounds. Distillates are produced in a variety of boiling ranges depending upon their intended use. They may be additionally processed to remove aromatic compounds; such products are called de-aromatized distillates. Distillates are used in a wide variety of products including camp, home heating, diesel, and jet fuels; charcoal starters; paint thinners; mineral spirits; and paint, stain, and insecticide vehicles.

Isoparaffinic products, as the name implies, consist of isoalkanes (isoparaffin) compounds. Like distillates, isoparaffinic products are produced in a variety of boiling ranges. Isoparaffins are marketed in an assortment of low-odor solvents, paint thinners, charcoal starters, manufacturing process fluids, polishes, paint, stain, and varnish transfer solvents, lubricants, and aviation fuels.

Naphthenic–isoparaffinic products, also called naphthenic–paraffinic products are petroleum distillate products with the normal alkanes and the aromatic compounds removed leaving isoalkane and cycloalkane (naphthenic) compounds. The uses of naphthenic–isoparaffinic products include low-odor solvents, paint thinners, charcoal starters, insecticide vehicles, and lamp oils.

Aromatic solvents are comprised solely of the aromatic compounds represented in crude oil within a given boiling range. The presence and abundance of alkylbenzenes, PNAs, and indane compounds is dependent on the boiling range and composition of the original crude. Aromatic solvents have high solvency powers and thus their uses include industrial cleaners, solvents, and degreasers.

Normal alkane products are very simple mixtures of normal alkanes, generally encompassing a boiling range of three to five compounds. Normal alkane products have a variety of uses including microencapsulation products used for NCR paper, solvents for waxes and polishes, insecticides, liquid candle fuels, lubricants, and agricultural chemicals (pesticides and herbicides) [4,5].

Nonpetroleum ignitable liquids

Nonpetroleum ignitable liquids represent a very small fraction of ignitable liquids, which is fortunate as they tend to be more difficult to recognize and identify. The two major subclassifications are oxygenated solvents and natural products extracted from plants.

Oxygenated solvents can be single compound products or complex mixtures. Their distinguishing characteristic is the presence of significant quantities of an oxygenated compound. Examples of common single compound products include acetone, ethanol, and methylethylketone (MEK). Some products, including lacquer thinners and enamel reducers, often consist of one or more oxygenated compounds blended with other chemicals (including petroleum products).

The most common naturally derived ignitable liquids are turpentine and citrus oil extracts. Turpentine is extracted from coniferous woods (soft woods) and is comprised of compounds called terpenes. Citrus oil extracts are extracted from citrus fruit peals. The major component of citrus oil extracts is *d*-limonene. *d*-Limonene is found in a wide variety of products including fragrances, cleaners, and flavorings.

Brand and product identification

With the exception of gasoline, the products of each classification typically have a number of uses, which generally makes product and/or brand identification impossible. Even gasoline, although produced for use as an engine fuel, is often used by consumers in other applications, notably as a solvent or degreaser. Numerous products and numerous brands of the same product may have identical formulations. Conversely, different lots of the same product and brand can have distinctly different compositions. Thus, ignitable liquids and their residues can typically be identified only in terms of classification.

Instrumental analysis

Analytical process

Ignitable liquid identification is made by comparison of samples to reference ignitable liquids. Through chemical and instrumental analysis, the properties of reference liquids, including ignitablity, solubility, boiling point range, and chemical composition can be established. By direct comparison of an unknown liquid or sample extract to such references, ignitable liquids and their residues can be identified.

While there are wide differences in crude oils, within distinct boiling ranges and functional groups they are predictably similar. Specific combinations of compounds, which ultimately result in specific chromatographic patterns of peaks are consistent and identifiable and are the basis of ignitable liquid recognition, identification, and classification by a process known as pattern recognition.

Pattern recognition is identification based, not on the presence of individual compounds, but rather, on the presence of compounds in relation to other compounds (Figure 6.1). Petroleum products are generally comprised of numerous (often hundreds) of compounds. These compounds are not unique to ignitable liquids. Many are common matrix contaminants and many are generated in the pyrolysis or incomplete combustion of various materials [6,7].

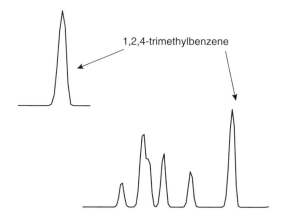

Figure 6.1 Pattern recognition: the presence of 1,2,4-trimethylbenzene (1,2,4 TMB) by itself is fire debris chromatographic data is not significant, however, when it is present as a portion of a key five-peak pattern as shown above, a petroleum product is indicated

Thus, the presence of any given compound in fire debris is not, in itself, typically significant. Fortunately, unlike in crude oil, the relative ratios of hydrocarbons produced by pyrolysis are generally inconsistent. By looking for specific patterns representing combinations of compounds, ignitable liquids can be differentiated from pyrolyzates or incomplete combustion products. These patterns are visualized using gas chromatographic (GC) techniques.

Gas chromatography

Gas chromatography is an instrumental process used for the separation and detection of volatile compounds. Using GC, fire debris extracts, as described in Chapter 5, are separated into components and their relative abundances are measured. Separation occurs by partitioning the analyte between a gas mobile phase and a liquid-coated stationary phase. A graph is produced, called a chromatogram, which depicts the retention time of each compound (the amount of time the compound in retained in the column) versus the detector response in terms of abundance (Figure 6.2).

A gas chromatograph consists of six basic parts: a carrier gas, an injection port, an oven, a column, a detector, and a data-recording device (Figure 6.3). The carrier gas (helium, hydrogen, or nitrogen) acts as the mobile phase to push the sample through the column to the detector. The sample is volatilized at high temperatures in the injection port where it is swept into the column by the carrier gas. The column, typically a thin capillary column coated with a bonded stationary phase liquid, will separate the sample based on the individual component's affinity to the stationary phase in a given set of conditions (carrier gas flow rate, oven temperature, etc.). Generally nonpolar columns are employed for ignitable liquid analysis due to the nonpolar nature of the hydrocarbons common to petroleum products. When using a nonpolar column, hydrocarbon separation is based on boiling points of the individual compounds.

The oven serves as the temperature regulator for the GC. Because the components of ignitable liquids tend to have broad boiling point ranges isothermal (single temperature) programs are not generally effective for producing quality results, thus temperature

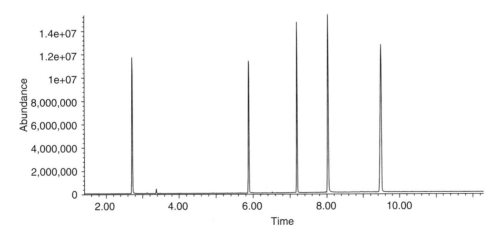

Figure 6.2 Data showing the separation of a five-component mixture using GC

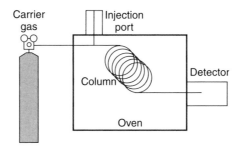

Figure 6.3 Schematic drawing of a typical gas chromatograph

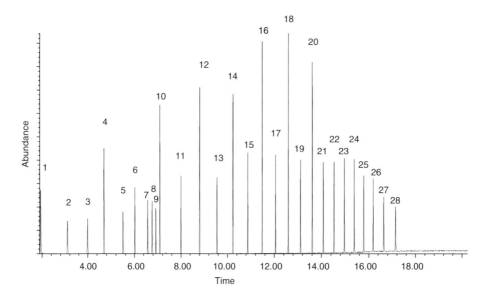

Figure 6.4 Gas chromatographic data showing the separation of *n*-alkane and aromatic compounds. 1 = hexane (C6), 2 = heptane (C7), 3 = toluene, 4 = octane (C8), 5 = *p*-xylene, 6 = nonane (C9), 7 = *m*-ethyltoluene, 8 = *o*-methyltoluene, 9 = 1,2,4,-trimethylbenzene, 10 = decane (C10), 11–28 = C11–C28, respectively. Acquisition parameters: HP 5890GC-5970MSD, 25 m DB-1 column with 0.2 mm ID, 0.33 μm film thickness. Injector port temperature: 250°C, detector (transfer line) temperature 280°C. Initial temperature: 45°C, initial time 3.0 min, ramp 20°C/min, final temperature 300°C, final time 4.5 min. Scan 33–400 *m/z*

programs are used which ramp the oven temperature over a given period of time. The oven can be programmed to optimize temperature conditions to allow for the best separation with greatest efficiency, most reproducibility, and appearance of best chromatographic data.

As a general rule, a chromatographic method (column, temperature program, etc.) used for routine ignitable liquid analysis should effectively separate and detect the normal alkanes from C6 through C20 and a variety of closely eluting aromatic compounds (Figure 6.4). Such a system allows for reasonable detection and separation of the most common ignitable liquid products. Additionally, a system that allows for the detection and separation of lower boiling compounds (i.e. methanol, ethanol, acetone, pentane) should be available.

The detector collects and measures the relative abundance of each component as it elutes from the column. There are two types of detectors that are suitable for ignitable liquid analysis: the flame ionization detector (FID) and the mass spectrometer (MS). The FID works by combusting the eluant from the column in a flame producing ions that are converted into a current. The resulting current is amplified, measured, and recorded.

The relative concentration of a single compound between several injections can be calculated, however, peak heights do not necessarily represent the actual concentration of one compound in relation to another in the same sample. Thus, data interpretation is not based on the concentration of one compound to another, but rather on the detector response of one compound to another.

Mass spectrometry

The most popular instrument used for ignitable liquid analysis is the gas chromato-graph–mass spectrometer (GC–MS). The mass spectrometer, also called the mass selective detector (MSD), is used in conjunction with the gas chromatograph to provide additional information regarding the compounds eluting from the column. In addition to the total ion chromatogram (TIC), which is comparable to the data obtained from the GC–FID, the mass spectrometer provides additional information allowing the analyst to classify and often identify the eluting compounds.

In GC–MS, compounds elute from GC column into the mass spectrometer where they are bombarded by an electronic beam that fragments the compound into ions based on chemical structure. These fragments are measured and recorded producing spectral data with peaks defined by fragment abundance (Figure 6.5). Fragmentation is predictable and, in most cases, unique to a given compound. Unfortunately, because hydrocarbons are very simple compounds and because chromatographic peaks may not be completely resolved, often mass spectral data obtained in ignitable liquid analysis is more chemical class specific

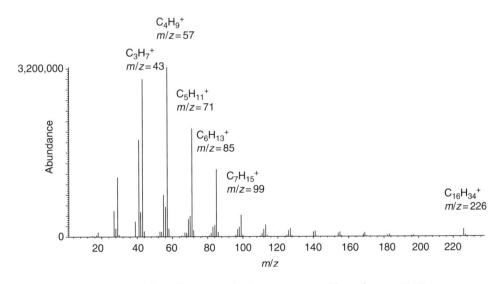

Figure 6.5 Mass spectral data illustrating the fragmentation of hexadecane (C16)

(i.e. alkanes, cycloalkanes, aromatics, etc.) than compound specific. For purposes of pattern recognition, this is more than sufficient.

The most important feature of the GC–MS, in terms of fire debris analysis, is the additional levels of pattern recognition that it can provide. Because each class of hydrocarbons has specific ions that are indicative of chemical class, ions can be extracted to create classes indicative chromatograms. For example, normal and isoalkanes consistently produce ion fragments $m/z = 43, 57, 71, 85$ (Figure 6.6). Extracting these ions from the total ion chromatograph produces an extracted ion chromatogram (EIC) indicative of the normal and isoalkanes in the sample (Figure 6.7). This strategy can be applied to each class of hydrocarbons (alkanes, cycloalkanes, aromatics, indanes, and PNAs) to produce less complex patterns for comparison to reference ignitable liquids [8]. The resultant EICs can simplify the data making ignitable liquid recognition and classification much more efficient.

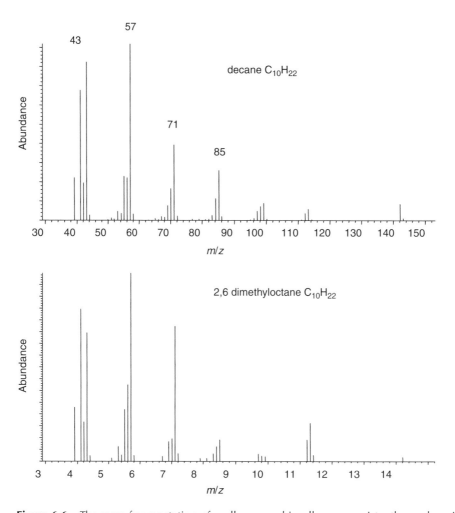

Figure 6.6 The mass fragmentation of *n*-alkanes and isoalkanes consistently produce ion 43, 55, 71, 85 ... Extraction of these ions from a total ion chromatogram will result in data indicative of the alkane content

164

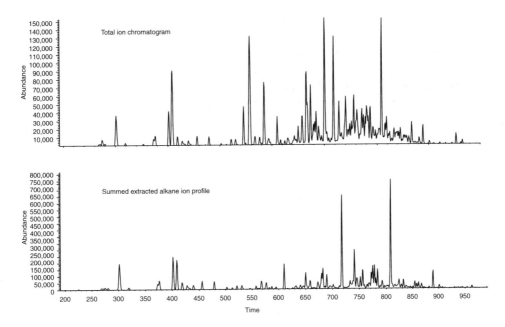

Figure 6.7 TIC of a gasoline and medium petroleum distillate mixture (top). Extracting the alkane ion profile (summed ions 43, 57, 71, and 85) results in a less complex chromatogram highlighting the alkane content of the mixture (bottom)

EIC has a filtering effect on the data reducing comparisons in highly complex GC patterns to less complex patterns representing key chemical classes (Figure 6.8). In samples with complex matrix contribution to the sample TIC, EIC can minimize chromatographic interference from the sample matrix. Additionally, the EIC data can be used diagnostically to help readily classify ignitable liquids and make more precise comparisons among samples and references.

Each class of hydrocarbons has characteristic ions based on successive alkyl (CH_2) substitution of the base compound, thus resulting in ion fragment increases in increments of fourteen. Appropriate ions for the various classes of hydrocarbons can be found in Table 6.2 [3,8,9]. It is important to note that ions are class indicative but not class exclusive. While all isoalkanes and normal alkanes of sufficient molecular weight will fragment to contain abundant ions 43, 57, etc., these compounds also contain minor ions that are indicative of alkenes 55, 69, etc. Thus the extraction of ion 55 does not result in the exclusive extraction of alkene compounds. This is true of most of the extracted ions used in fire debris data interpretation.

EICs may be produced by singularly extracting and evaluating individual ions or by adding class-specific ions for evaluation. Single ion EIC provide more discrete data. In the case of aromatic compounds, single ion EIC can result in individual chromatograms highlighting each alkyl-substituted grouping, that is, C2, C3, and C4 alkylbenzenes. Summed EICs provide data with increased signal/noise ratios (i.e. increased overall sensitivity) and direct component abundance comparisons within the ion class (Figure 6.9). Summed ion chromatograms are generally preferred as they provide more obvious information regarding relative ratios of similar class compounds throughout the boiling range of the sample, that is, the relative abundances of C1, C2, C3, and C4 alkylbenzenes in relation to each. Additionally,

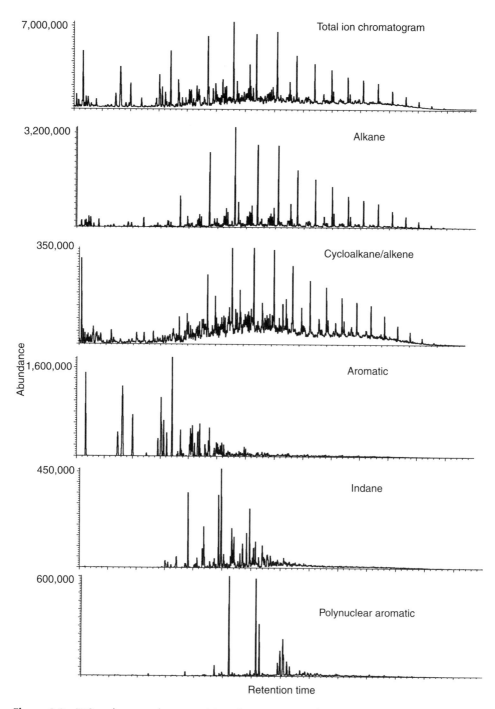

Figure 6.8 TIC and summed extracted ion chromatograms from a standard accelerate mixture (SAM) consisting of 50% evoporated gasoline, kerosene, and diesel fuel (1:1:1)

Table 6.2 Chemical composition of various classes of hydrocarbons

Petroleum class	Chemical composition
Gasoline	Highly aromatic, wide boiling range (C4–C12), less abundant aliphatic compounds present.
Aromatic solvents	Exclusively aromatic.
Distillates	Highly aliphatic with predominant normal alkanes, aromatic content varies.
Isoparaffinic products	Exclusively isoalkanes compounds.
Naphthenic–paraffinic products	Exclusively aliphatic with predominate isoalkane and cycloalkane compounds.
Normal alkane products	Exclusively normal alkanes, typically spanning 3–5 consecutive compounds.

summed EIC require far fewer chromatograms to obtain much of the same information that would be obtained using single ion methods. Both summed and single ions techniques are appropriate as long as the analyst is fully aware of the interpretation and limitations of each. There are instances where the matrix contaminants contribute interfering peaks in the summed EIC data where single ion EIC can provide more helpful information [10].

When using EIC for data interpretation it is important to evaluate data from each of the chemical classes as many of the ignitable liquid classes have similar EIC patterns. While chromatographic patterns are important and diagnostic, the relative abundance of EICs to that of different chemical classes is very important. For example, refinery products that contain aromatic components (gasoline, aromatic solvents, distillates) will have comparable, and readily identifiable, aromatic profiles within a given boiling range (Figure 6.10). Relying simply on the aromatic pattern, a medium petroleum distillate can easily be misidentified as gasoline (Figure 6.11). However, comparing the relative abundance of the aromatic content to that of the aliphatic content, there is an obvious predominance of aliphatic compounds in the distillate, the reverse is true in gasoline.

Because of variants in raw crude oil, the relative ratios of one chemical class to another can vary somewhat. Databases of a wide variety of exemplars for local markets' various products in all the ignitable liquid classes should be maintained to establish acceptable deviations in ratios [11].

GC–MS data can be collected in one of two modes: scan or selected ion monitoring (SIM). In the scan mode all of the ions are collected and measured within a given range (10–400 amu is common for ignitable liquid analysis). In the scan configuration, all the ion fragments in the range are collected by the detector and indicated in the resultant data. EICs are typically generated from scan mode data. In SIM, only pre-selected ions are measured and recorded. Because the detector has fewer ions to measure, the sensitivity is greatly increased. In fire debris analysis the same ions that are used in extracted ion chromatography are typically used in SIM (Table 6.2). Non-hydrocarbon sample matrix interferences are minimized resulting in cleaner data. At first glance it would appear that the use of SIM would be preferable to scan for ignitable liquid analysis, however, this is not the case. Although the sample matrix can contribute a significant amount of interfering compounds, often the presence of these compounds is critical to the proper interpretation of the data.

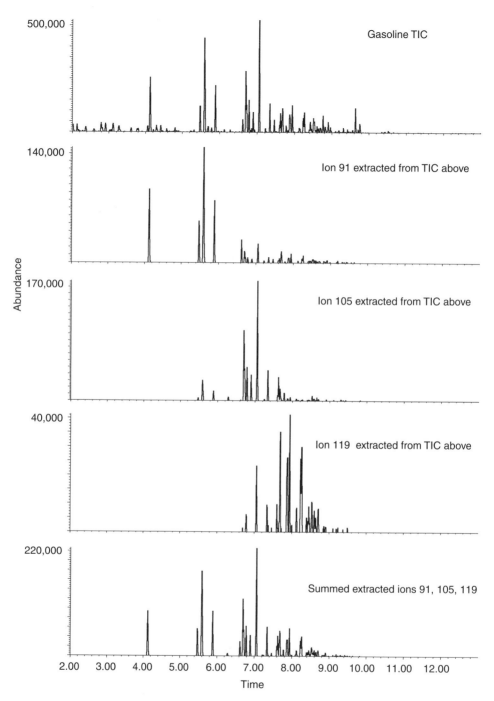

Figure 6.9 Single and summed extracted aromatic indicative ions from a partially evoporated gasoline sample

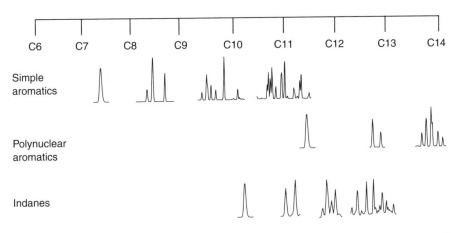

Figure 6.10 Aromatic patterns found in various petroleum products within given boiling ranges. Note that the pattern and the elution order are based upon GC separation using a nonpolar column

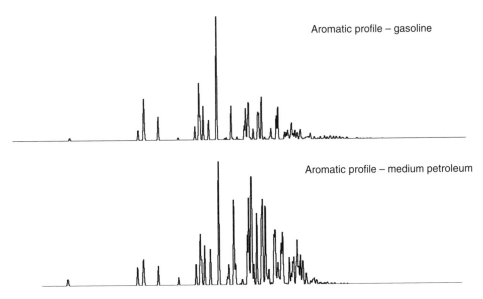

Figure 6.11 Aromatic profiles (summed ions 91, 105, and 119) of a partially evaporated gasoline and a medium petroleum distillate are virtually identical. Distinguishment between ignitable liquids must include both the individual EIC presentation and their relative abundances. In this example, while these products contain similar aromatic compounds, in gasoline the aromatics are much more abundant than the alkanes; in the distillate, the converse applies to the distillate

Thus, SIM is generally reserved for analyses when scan data indicates the presence of an ignitable liquid but lacks the level of sensitivity to make a definitive identification.

All data used for comparison (sample, reference ignitable liquid, and comparison samples) should be generated using similar techniques. Summed ions used for sample data extraction should be compared to the same summed ions extracted from the reference data.

169

SIM sample data should be compared to SIM reference data. Ideally, both the sample and the reference should be run on the same GC system using identical instrumental conditions. Fortunately, with the use of computer programs (methods and macros) the data acquisition and ion extraction processes can be automated [11,12].

Ignitable liquid classification

Data interpretation

As previously stated, there are two broad classes of ignitable liquids (petroleum and nonpetroleum) and nine general classifications. The petroleum-based classifications are gasoline, petroleum distillates, de-aromatized petroleum distillates, isoparaffinic products, naphthenic–paraffinic products, aromatic solvents, and normal alkane products. The nonpetroleum-based classifications are oxygenated solvents and a catchall miscellaneous class that includes natural products.

Each class of ignitable liquids has specific chemical properties resulting in diagnostic chromatographic data allowing for recognition and identification. Within a given boiling point range, functional group compounds are very reproducible in terms of relative concentrations. The more complex the product, that is, the greater number of components, the easier recognition and identification become. The diagnostic features of each class are provided in the subsections below and are summarized in Table 6.3.

Aside from chemical composition, petroleum products can be further classified based on their boiling point range. Most petroleum-based ignitable liquids fall into one of three distinct boiling point ranges simply titled light, medium, and heavy [13]. Light petroleum products generally elute before C9 (nonane). Medium petroleum products are typically of narrow boiling range – spanning three to four normal alkanes – and typically center around C9, C10, or C11. Heavy petroleum products have a broader boiling range, spanning five or more normal alkanes and typically center above C11.

Gasoline

Gasoline is a blended refinery product that elutes in the range C4–C12 with predominant aromatic compounds. Most gasoline contains aliphatic compounds, however, these alkanes are significantly less abundant than aromatics. Aromatic compounds result in increased octane ratings where aliphatic compounds generally lower them. The composition of the

Table 6.3 Class indicative ion fragments used for EIC in hydrocarbon analysis

Hydrocarbon class	Ions				
Alkanes (iso and normal)	43	57	71	85	99
Cycloalkanes and alkenes	55	69	83	97	—
Simple aromatics	91	105	119	133	—
Polynuclear aromatics	128	142	156	—	—
Indanes	117	118	131	132	—

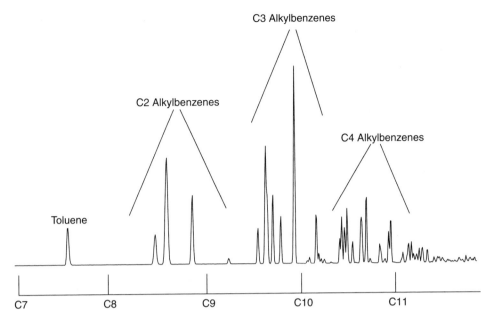

Figure 6.12 Aromatic extracted ion profile of gasoline. The C3 and C4 alkylbenzene patterns are considered the most diagnostic for determining the presence of aromatic bearing petroleum-based ignitable liquids

aromatic pattern associated with gasoline is highly consistent and easily recognizable. The aliphatic composition varies greatly based on crude oil composition and refinery process streams utilized in manufacture.

The major components in gasoline are the simple aromatics (alky-substituted benzenes). The most diagnostic in terms of pattern recognition are the five-peak C3 alkylbenzene configuration and the C4 alkylbenzene series of peak doublets (Figure 6.12). These aromatic peak patterns can be seen readily on both the full scan (TIC) or GC–FID data and on the aromatic extracted ion profile. Other consistent chromatographic features include the indane and PNAs. A diagnostic methylindane doublet elutes around C11 (on a nonpolar column), because of the concentration and co-elution in that region, indane patterns are only of benefit when using EIC. The presence of naphthalene and the methylnaphthalene doublet is dependent on gasoline formulation. Cold weather markets may not contain these compounds. When they are present, the naphthalene and methylnaphthalene peak grouping, which elute between C11 and C13, is generally consistent (Figure 6.13). The aliphatic peak patterns in gasoline vary widely and are not as significant in appearance as it is in presence. The absence of aliphatic compounds would indicate the presence of an aromatic solvent rather than gasoline.

Petroleum distillates

Distillates have very high aliphatic content with predominant normal alkanes. The normal alkanes are represented as spiking peaks in a normal distribution in medium and heavy distillates. In light range distillates, the *n*-alkanes tend to be less predominant, but are still significant (Figure 6.14). The boiling range will vary based on intended use, however, within

171

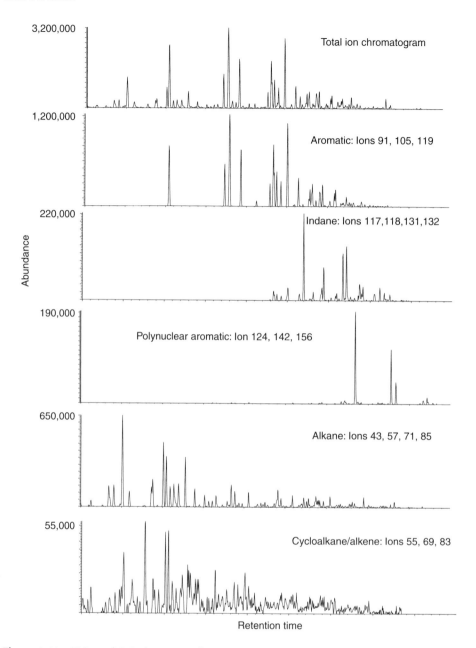

Figure 6.13 TICs and EICs for a partially evaporated gasoline. Note the different ion abundances of the different classes

a given boiling range, GC patterns for both major and minor components and chemical classes are consistent.

Heavy petroleum distillates that include the C17–C18 range also contain the compounds pristane and phytane. The resulting doublets (C17 followed by pristane, and C18 followed by phytane) are easily recognizable diagnostic features of heavy petroleum distillates

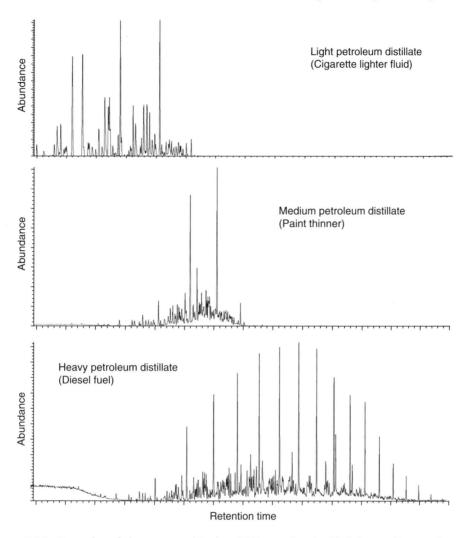

Figure 6.14 Examples of chromatographic data (TIC) associated with light, medium, and heavy petroleum distillates. Note the predominant spiking normal alkane peaks

(Figure 6.15). In order to identify an unknown as a distillate encompassing the C17–C18 range, pristane and phytane must be present. The absence or these compounds in the presence of a homologous series of spiking normal alkanes is indicative of the presence of polymer pyrolysis rather than petroleum products – see "Matrix Contributions."

The concentration of the aromatic constituent of distillates can vary greatly based both on natural variants of crude oil [1] and additional product processing. Generally, aromatics will be present in the common distribution but at levels significantly less abundant than the aliphatic compounds. De-aromatized distillates are refined to remove or diminish toxic aromatics for low-odor products like charcoal starters and various solvents. Thus, the absence of significant aromatic compounds is an indicator of a de-aromatized product. Conversely, some product manufacturers blend in aromatic fractions to meet product specifications, thus

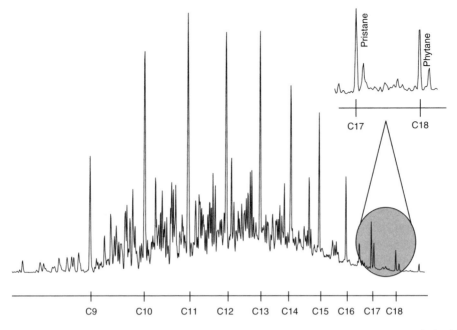

Figure 6.15 TIC of a heavy petroleum distillate highlighting the presence of peak doublets at C17 and C18

resulting in aromatic compounds in increased ratios. In some cases, it may not be possible to differentiate de-aromatized, traditional, and re-aromatized distillates using GC–FID alone. EIC will allow for easy recognition of de-aromatized products (by the diminishment or absence of aromatic compounds in the extracted ion profiles). EIC may allow for the recognition of distillate/aromatic blends depending on the boiling range and concentration of aromatic solvent additives. The analyst must recognize that natural variants in aromatic content in distillate fractions is common and allow that distinguishment between traditional distillates and processed blends may not always be possible, nor is it typically necessary.

When using EIC in data interpretation, attention must be given to relative peak abundance both within and between the various classes of compounds. Differentiating a distillate from gasoline using GC–FID or TIC data is fairly obvious. However, EIC can result in data that may lead one to the false conclusion that mixtures are present, when in fact, the ion profiles in distillates are often similar in composition, but significantly different in relative abundances, than that of gasoline (Figure 6.16).

Isoparaffinic products

Isoparaffinic products are comprised solely of isoparaffinic compounds. Isoparaffinic products have distinctive characteristics but unfortunately lack easily recognizable diagnostic peak patterns in their total ion chromatographic data. Isoparaffinic products in the light to medium ranges generally result in distinct patterns with fairly well-resolved peaks. As the boiling points increase, the number of potential isoalkane isomers increases resulting in increasing complicated and less resolved patterns (Figure 6.17).

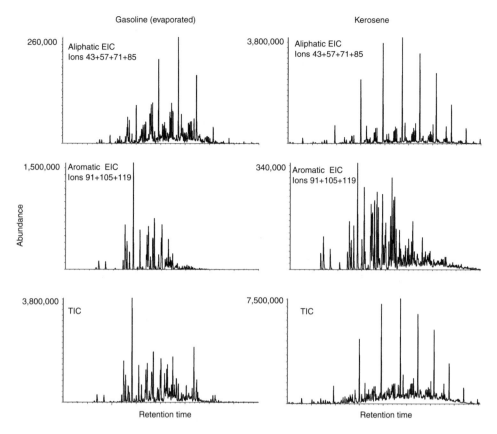

Figure 6.16 Chromatographic data representing a 90% evaporated gasoline (left) and kerosene, a heavy petroleum distillate (right). While the extracted ion data presentation is similar, the relative abundance of aliphatics to aromatics is significantly different

Using extracted ion chromatography, isoparaffinic products are much easier to recognize as the TIC, alkane, and cycloalkane extracted ion profiles have very similar patterns, but in decreasing order of abundance. This is because peaks represented in each chromatogram (TIC, alkane EIC, and cycloalkane EIC) represent the same compound. The alkane and cycloalkane indicative ions are represented in each compound in consistent proportions. Ions 55, 69, and 83, which are diagnostic for alkenes and cycloalkanes, are also minor ions in both normal and isoalkanes. Thus, the extraction of the "cycloalkane/alkene" indicative ions from isoparaffinic products results in significantly less abundant, but otherwise identical patterns to that of the isoalkanes ions (Figure 6.18).

Naphthenic–isoparaffinic products

Naphthenic–isoparaffinic products, also called naphthenic–paraffinic products predominantly consist of isoalkanes and cycloalkanes. *n*-Alkanes may be present in some formulations, however, in greatly reduced amounts compared to petroleum distillates. Aromatic compounds are not present. Most common naphthenic–isoparaffinic products fall

175

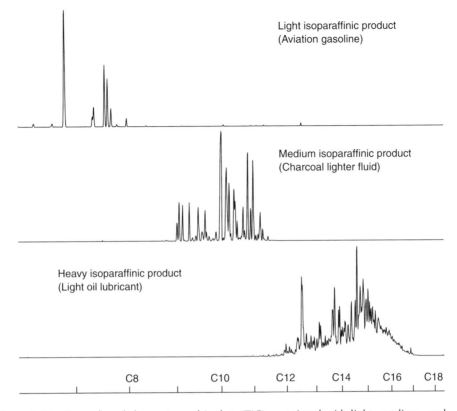

Figure 6.17 Examples of chromatographic data (TIC) associated with light, medium, and heavy isoparaffinic products. Note as the boiling range increases the peaks become less resolved

into the medium or heavy range. Their data are characterized by an unresolved envelope of peaks. Both recognition and identification using GC–FID or TIC alone can be difficult due to the lack of distinctive diagnostic peak patterns. The EIC data is exemplified by abundant alkane and cycloalkane extracted ion profiles. Unlike isoparaffinic products, the presentation of the alkane and cycloalkane are distinctly different in naphthenic–isoparaffinic products (Figure 6.19).

Aromatic solvents

Aromatic solvents, as the name implies, exclusively contain aromatic compounds. Because gasoline is high in aromatic content, aromatic solvents that elute in the C7–C12 range can easily be confused with gasoline at first glance, especially in the presence of matrix interferences. In neat samples, aromatic solvents generally result in much cleaner, better resolved, GC data due to the lack of aliphatic constituents that are found in gasoline. Most aromatic solvents tend to encompass distinct narrow boiling ranges (Figure 6.20). Aromatic solvents are marketed primarily for their high solvency powers and are often used in water-based products (emulsions) including insecticides, thus considerations should be made in data interpretation for non-ignitable product sources of aromatic solvents found in debris.

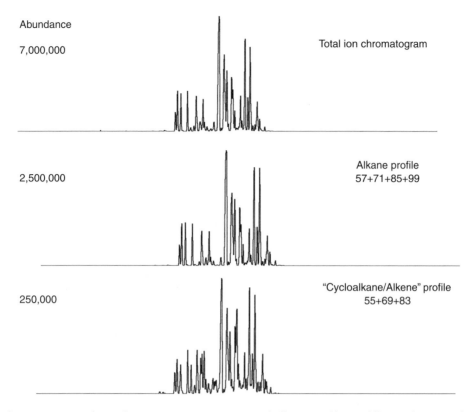

Figure 6.18 Total ion chromatogram (top), extracted alkane profile (middle), and extracted "cycloalkane/alkene" profile (bottom) for a medium-range isoparaffinic product. Note that all the profiles look identical. The "cycloalkane/alkene" ions 55, 69, and 83 are minor ions of alkane compounds (there are no cycloalkane or alkene compunds present) and thus are represented in the "cycloalkane/alkene" profile as they are represented in the alkane profile. The relative abundances of the classes as presented differ by a ratio of 10 : 1

Caution should be exercised when evaluating light and in some cases, medium, aromatic compounds in debris samples as these compounds are produced, often in common ratios, in the pyrolysis of some common polymer materials [6]. The comparison of the extracted ion chromatographic peak patterns of indane and, in some cases and depending on the matrix composition, PNAs can be a valuable tool for differentiating medium-range aromatic products from sample matrices.

Normal alkane products

Normal alkane products are unusual petroleum products because they contain so few compounds. *n*-Alkane products are generally comprised of 3–5 normal alkanes (Figure 6.21). Pattern recognition is not a viable technique for identification (although it can be extremely helpful in screening) because there are not enough points of comparison to be meaningful. Additional evaluation including GC retention time comparison and mass spectral interpretation is recommended.

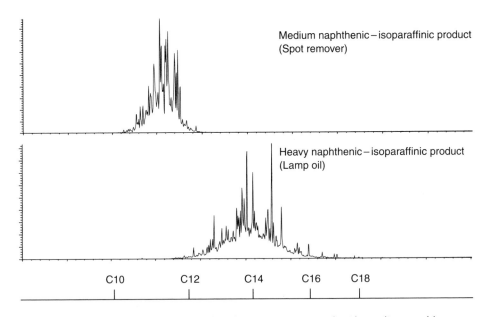

Figure 6.19 Examples of chromatographic data (TIC) associated with medium and heavy naphthenic–isoparaffinic products. Note the highly unresolved peak grouping

Oxygenated solvents

Pattern recognition is not typically a reliable technique for the recognition and identification of most nonpetroleum-based ignitable liquids. These products are often comprised of few compounds, many of which may be formed by matrix decomposition, or are naturally occurring and thus incidental to many substrates.

Oxygenated solvents, as the name implies, are ignitable liquids that contain oxygenated compounds, most commonly low molecular weight alcohols and ketones. The identification of an oxygenated compound found in debris as an ignitable liquid and thus a potential accelerant should be done with caution. Many common oxygenated compounds are produced in the incomplete combustion of common materials, so their presence in fire debris extracts should be expected [14]. The ASTM test method for analysis of ignitable liquid residues recommends a minimum abundance of at least one order of magnitude above other matrix compounds before the finding of a single low molecular weight oxygenated compound should be considered significant [13].

Most of the common oxygenates of interest elute prior to C6 on nonpolar GC columns and thus may not be sufficiently resolved using methods optimized for petroleum-based products. Alternative GC methods including the use of more polar columns, thicker column phases, and/or alternative temperature programs may be required for detection and identification. Because many are not components of complex mixtures, pattern recognition techniques are not suitable for identification. GC retention times coupled with MS identification is the most common and reliable analytical practice.

Oxygenates may be added to petroleum-based ignitable liquids in some specialty products, notably in lacquer thinners and enamel reduces (Figure 6.22). In these instances, a combination of component identification and pattern recognition may be required.

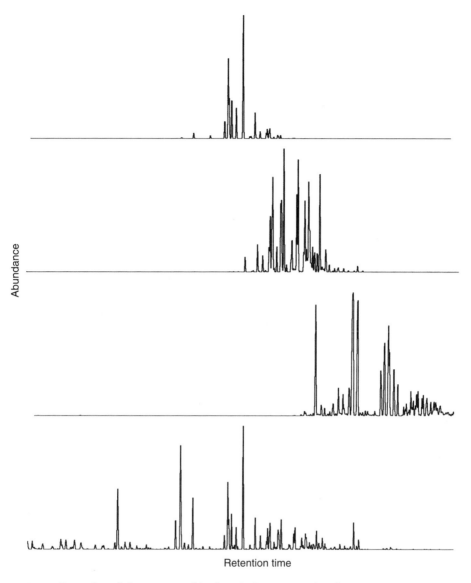

Figure 6.20 Examples of chromatographic data (TIC) associated with some common aromatic solvent products and gasoline. Note that the boiling range of the aromatic solvents is generally more narrow and cleaner than that of gasoline (bottom)

Natural extracts

Other nonpetroleum derived ignitable liquids include those extracted from natural products. Turpentine is extracted from coniferous (soft) woods for use in solvents, fragrances, insecticides, and finishing oils. Citrus oil extracts are extracted from citrus fruits. Both contain terpenes. Terpenes are unsaturated hydrocarbons found in essential oils and oleoresins of plants. Turpentine consists of a variety of terpene compounds with α- and β-pinene typically being the most abundant. The predominant component of citrus oil extracts is the

179

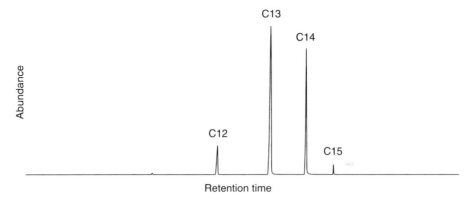

Figure 6.21 Examples of chromatographic data (TIC) associated with a common normal alkane product (candle oil)

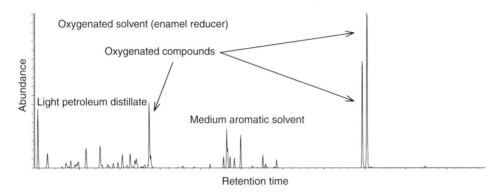

Figure 6.22 Examples of chromatographic data (TIC) associated with an oxygenated solvent (enamel reducer). Oxygenates are often found in mixtures with petroleum-based ignitable liquid as shown above

terpene *d*-limonene. Identification of terpenes in debris extracts is very straightforward with MS identification being the most common practice.

The challenge of terpene analysis is not the identification but rather the assessment of relevance of the identification to the fire investigation. A large percentage of fire debris samples include soft woods, an obvious potential source for the terpenes found in the GC data. The argument can be made that in the absence of visible/recognizable wood in the debris, a conclusion that includes turpentine as a potential accelerant would be viable. Unfortunately, with the amount of wood used in common structures the proximity of any matrix with terpene-laden wood in the presence of time and heat can logically result in transference of identifiable amounts of terpenes. Additionally, there are a great number of cleaning products and solvents that contain terpenes thus making matrix composition of a single sample a somewhat dubious determinant of terpene significance.

Therefore, the identification of turpentine as a potential ignitable product in a given sample is generally restricted to samples with a notable absence of incidental sources, a lack of presence in credible comparison samples, abundances greatly beyond normal expectations for the given matrix, and/or the other mitigating factors. In any situation, when turpentine is reported as a potential product in debris samples, disclaimers to natural sources are recommended [15,16,17]. *d*-Limonene has incredibly widespread use in a wide variety of products and processes. It can be found in most "citrus" fragrant products [18]. The significance of *d*-limonene in debris is dependent on matrix composition, function and history; location; presence in multiple unrelated (different matrix) samples; presence in credible comparison samples; and relative abundance.

Evaporation effects

Because most ignitable liquids contain hundreds of compounds in relatively broad boiling ranges, the effect of environmental stressors (exposure, heat) can result in notable changes in the chromatographic pattern. More volatile compounds evaporate at a higher rate than less volatile compounds. This evaporation, also called "weathering," results in a shift of the chromatographic pattern towards higher boiling compounds. The impact of weathering on ignitable liquid composition, and thus data presentation, is dependent on the boiling range of the product and the amount of evaporation.

The GC data for fresh gasoline is significantly different from that of 90% evaporated gasoline. In the more evaporated sample, the lighter components are diminished resulting in a greater representation of the higher boiling point components (Figure 6.23). Within a given boiling range (e.g. C11–C12) the relative ratios of the compounds remain similar and products are still identifiable using pattern recognition techniques. The effect of evaporation is less dramatic for ignitable liquids with narrower boiling point ranges.

Evaporation will also affect the relative abundances of the extracted ion profiles. In fresh gasoline the aromatic to aliphatic ion ratio for the ions extracted is approximately 10:1, at 90% evaporated, that ratio is roughly 1:1. The predominant aromatic compounds in gasoline elute early, as they evaporate, the class-specific ion ratios shift as expected. However, the ion ratios within narrow boiling point ranges remain fairly constant.

Classification summary

Table 6.2 describes the various petroleum-based ignitable liquid classes and their general chemical characteristics. Databases of various ignitable liquids have been established and can aid in sample comparison and ignitable liquid classification [11,19]. However, final identification should be made based on data obtained in the same manner, under the same chromatographic conditions, and preferably on the same instrument.

Sample matrix and data interpretation

Matrix contributions

Unfortunately, the data obtained from analysis of debris extracts is not limited to components of ignitable liquids. Volatile compounds present in extracts can include background contaminants, pyrolysis products, incomplete combustion products, and ignitable liquid constituents. These compounds are all represented in the GC data. This can result in highly complex patterns in which the analyst must discern any ignitable liquid components from matrix contribution.

When the concentration of the ignitable liquid is relatively high or the contribution from the matrix is minimal, data interpretation is fairly straightforward. However, this is often not

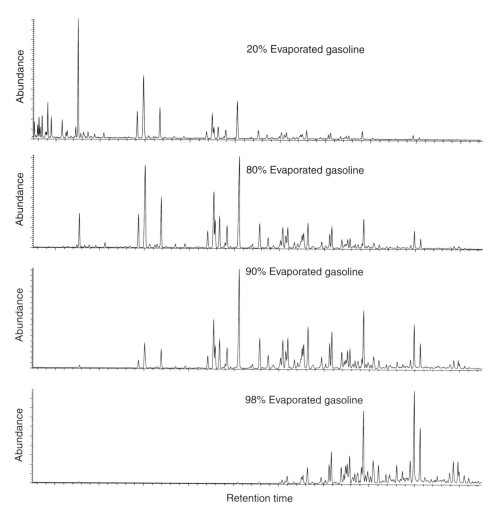

Figure 6.23 GC data of a progression of evaporated (weathered) gasoline. Note as the more volatile components evaporate, the less volatile become more distinct

the case. The amount and type of matrix contribution is dependent on composition, quantity, and history. Some matrices result in readily identifiable diagnostic patterns, others result in more random peaks.

Polyethylene plastics

The pyrolysis of polyethylene plastics results in data with some similarity to petroleum distillates. Most notably, both contain homologous series of spiking peaks that include the *n*-alkanes. The most notable difference is that in HDPE pyrolysis the homologous series is not of single peaks but of triplets representing corresponding alkadienes (aliphatic compounds with two double bonds), alkenes (aliphatic compounds with one double bond), and alkanes. Additionally, polyethylene plastic pyrolysis patterns lack the substructure, mainly isoalkanes and cycloalkanes including pristane and phytane, found in petroleum distillates (Figure 6.24).

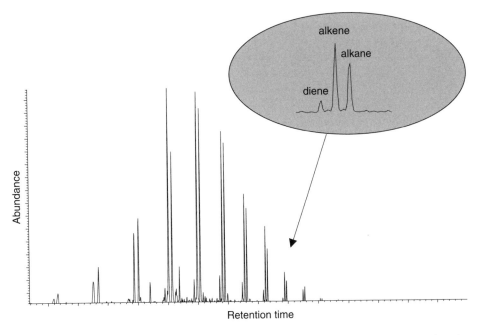

Figure 6.24 High-density polyethylene pyrolyzes to produce a homologous series of triple peak patterns representing consecutive alkyldiene, alkene, and alkane compounds

Carpet and carpet pad

Carpet and carpet underlayment pads are generally comprised of man-made fibers (nylon, polystyrene–butadiene, etc.). Some of pyrolysis products associated with these materials and that are commonly found in fire debris sample extracts include styrene, α-methylstyrene, and various simple and complex aromatics. Many of which are also found in petroleum-based ignitable liquids [6,7].

The extraction of aromatic ions from the resultant TIC can result in patterns with a great deal of similarity to gasoline and other aromatic-bearing petroleum products (Figure 6.25). In those situations, the most critical points of comparison for determining the significance of aromatic contributions to the GC pattern are the C4 alkylbenzene pattern, the indane pattern, the presence of styrene compounds, the abundance and presentation of aromatic peaks compared to matrix peaks, and pattern indications visible in the TIC. The C3 alkylbenzene patterns are less indicative (variances between petroleum products and matrix contribution can be minor) and the C2 are fairly useless. The PNAs are also less significant as they are generally produced as pyrolysis products, however they should not be completely excluded from the interpretation process. A distinct variance from the expected ratio of 2 and 1-methylnaphthene may be an indicator of matrix rather than petroleum product contribution.

This is not to imply that any compound or peak groupings found in ignitable liquids should not be considered, they must be present in proper ratios within the boiling point ranges, however, the possibility that some key patterns may be produced by the matrix must be realized and included into the data interpretation process.

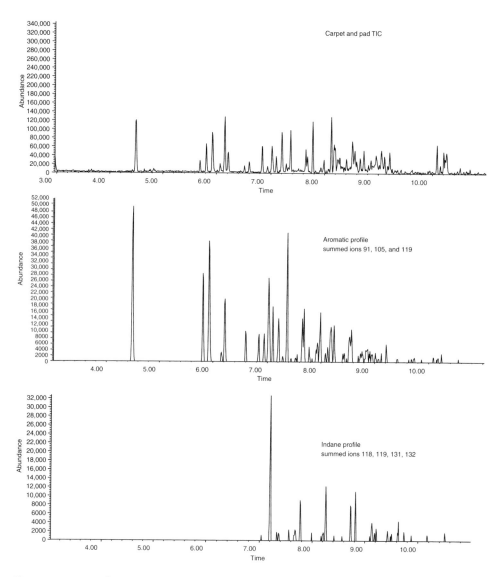

Figure 6.25 Simple aromatics are common pyrolysis products of some synthetic materials. The resulting EICs can have some similarity to petroleum-based products

Asphalt shingles

Asphalt shingles are common roofing materials and thus are common in fire debris samples. Analysis of the smoke condensates derived from burned asphalt shingles can result in chromatographic data with characteristics similar to heavy petroleum distillates, including the presence of pristane and phytane. They can be differentiated by the presence of alkenes in asphalt condensates, which are not present in distillates. The alkenes are not generally abundant and thus it is typically easier to differentiate petroleum distillates from asphalt condensates with the use of extracted ion chromatography.

The ions common to cycloalkanes are the same ions common to alkenes (note that both classes of compounds have the chemical formula C_nH_{2n}). Ion 55 is more indicative of alkenes than cycloalkanes and thus can be used to evaluate the presence of alkenes in the data. Extraction of ion 55 from GC data obtained by the extraction of asphalt smoke condensates will result in readily identifiable homologous series of doublets represented the *n*-alkene and *n*-alkane compounds (Figure 6.26). Ion 55 extraction from petroleum distillates results in a homologous series of single peaks [20].

Because these compounds arise from smoke condensate of asphalt rather than from the presence of solid material, the absence of roofing or other asphalt bearing materials in a given sample is not sufficient basis to conclude that asphalt is not contributing to the debris sample extract. Investigators should note the type of roofing material present at a structure

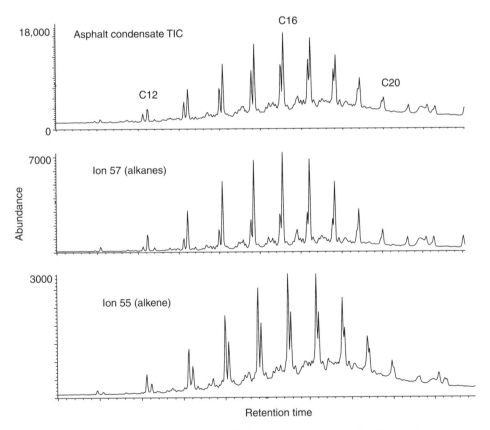

Figure 6.26 Chromatographic data from asphalt condensate sampled using passive headspace concentration on active carbon. The alkane data presentation (middle) is indicative of a petroleum distillate. The intense doublets in the alkene indicative data (bottom) are associated with the homologous alkene/alkane series. In the presence of heavily obscured TIC data, evaluation of both alkane and alkene indicative chromatograms is important to avoid false identification of a petroleum distillate (data courtesy of John Lentini)

fire and the analyst should routinely assess data for the presence of *n*-alkenes in apparent heavy petroleum distillate patterns.

Soil

Soil is a highly absorbent and insulating substrate that makes it an excellent matrix in terms of ignitable liquid retention [21]. Unfortunately, soils, notably fertile soils, can contain hydrocarbon-digesting bacteria that can significantly affect the composition. The resulting GC presentation of ignitable liquid extracts is often distorted and can be unidentifiable (Figure 6.27). Microbial degradation from soil can result in the loss of aliphatic, notably *n*-alkane, and/or aromatic compounds depending on the type of bacteria present and other environmental conditions [22,23]. This degradation can begin in a matter of a few days and can be so significant as to preclude the ignitable liquid identification.

Fortunately, soil samples do not typically contain matrix interfering compounds so the presence of significant hydrocarbons, even in grossly distorted patterns, can be an indicator of the presence of a petroleum product. The degradation in similar soils is relatively consistent, so by using comparison samples of soils spiked with plausible ignitable liquids, comparable data can be obtained, allowing for identification [22]. Avoiding degradation is obviously the best solution. Whenever possible, samples should be collected and analyzed quickly. Freezing samples immediately after collection and until sample extraction will retard degradation [22].

Incidental ignitable liquids

Ignitable liquids in general, and petroleum products in particular, are used in a wide array of industrial processes and everyday activities that result in incidental contamination of

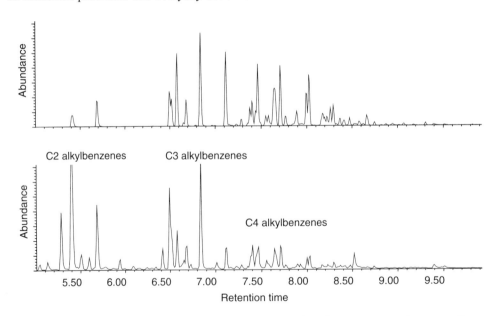

Figure 6.27 GC data of gasoline degraded in bacteria rich soil (top) compared to a gasoline reference ignitable liquid (bottom). Note the loss of key aromatic compounds

materials [23]. Environmental contaminants common in many materials can result in the detection of petroleum-based products. Extreme care should be exercised in identifying minute quantities of ignitable liquids in some matrices as potential accelerants, especially in the absence of a suitable comparison sample.

A comparison sample is "a sample of material collected from a fire scene which is, to the best of the investigator's knowledge, identical in every respect to a sample suspected to contain an accelerant, but which does not contain an accelerant" [24]. These are samples where the composition and probable history prior to the fire is similar to a suspect sample. Ideally, comparison samples for each sample matrix should be collected. Unfortunately, this is not practical and often not possible given the destruction level and the nature of the fire. In most instances, comparison samples are not needed, however in situations where ignitable liquids are identified, but probable incidental sources exist, comparison sample can be essential in establishing the significance of the analytical results.

The following are just a few of the many potential materials where incidental ignitable liquids are commonly found and why matrices, ignitable liquid concentrations, and sample history must be considered as a routine course in an investigation.

Shoes

Often suspect's clothing, including shoes, are taken in an effort to establish a link between any ignitable liquids found in scene samples and ignitable liquids found on the suspect's clothing. Unfortunately, petroleum products, most notably toluene and medium and heavy petroleum distillates are used in the manufacture of shoes and thus findings may lead to false conclusions. Shoes should always be packaged separately so that one can serve as a comparison sample, of sorts, for the other. When distinctly different ignitable liquids or ignitable liquid concentrations are found, the conclusion that a foreign ignitable liquid had been in contact with the shoes is credible. When the data obtained from the samples is similar the likelihood is that the ignitable liquid is incidental. Of course, this will depend on the type of shoe and the type of ignitable liquids present.

Insecticides

Commercial and retail insecticides commonly have petroleum-based ignitable liquids as a dispersement vehicle. Commercial insecticides often consist of medium-range aromatic solvents mixed with water to form an emulsion. Retail spray insecticides are often dissolved in medium and heavy distillates, isoparaffinic, or naphthenic–isoparaffinic solvents. The presence of these products on recently treated materials, notably carpets and baseboards, is not uncommon.

Finished Woods

Medium petroleum distillates and isoparaffinic products are common solvents for wood stains and finishes. Thus, the analysis of finished woods, including furniture and flooring, often results in the detection and identification of an ignitable liquid. Because wood can be highly absorbent, often these solvents can be readily detected years after the initial application. The incidental presence of these types of products in wood flooring must be considered whenever identification is made. The analysis of a comparison sample of like materials may be necessary to data interpretation [25].

Paper products

Some NCR paper ("no carbon required" – self-duplicating paper) is produced using microencapsulated *n*-alkane products. Kerosene and similar heavy petroleum products are often used as transfer vehicles for inks and thus are often found in the analysis of books, magazines, and newspapers. Thus, the identification of kerosene in a sample that contains newspaper is not exceptional. Without this knowledge an improper conclusion identifying the incidental kerosene as an accelerant may result.

Conclusion

A common thread throughout this chapter is that forensic laboratories analyze samples for the presence ignitable liquids, not accelerants. Investigators and analysts should then work together to determine the significance of those findings in relation to the circumstances of the overall investigation, most notably the scene investigation. The determination of potential accelerants will be based on a number of factors and will typically be unique to each investigation. Some of the factors that should be considered include the following.

- The presence of similar ignitable liquids in multiple samples.
- The relationship of positive samples to scene indicators, that is, pour patterns, fire origins, etc.
- The type of ignitable liquid identified in relation to the sample matrix.
- The type of ignitable liquid identified in relation to the type of scene, i.e. home, office, warehouse, vehicle, etc.
- The relative abundance of the ignitable liquid identified in relation to the sample matrix.
- The absence/presence of similar ignitable liquids in credible comparison samples.
- The type of ignitable liquid identified in relation to scene/sample history, i.e. recent pest control efforts, furniture re-finishing, etc.

Table 6.4 Some potential scenarios to explain laboratory findings

Reasons for positive laboratory findings	Reasons for negative laboratory findings
An accelerant was used and the fire was incendiary. An incidental ignitable liquid was present in the sample. The sample was contaminated with an ignitable liquid in the coarse of collection, packaging, or processing.	No accelerant was used in the fire. An accelerant was used but was entirely consumed by the fire. An accelerant was used but was not present in the samples collected. An accelerant was used and is present but was not identifiable due to concentration or matrix interference.

In no case should the determination of fire cause (i.e. arson, accidental, electrical, etc.) be based solely on the identification of an ignitable liquid. Given the preponderance of ignitable liquids, especially petroleum-based ignitable liquids, used in our environment, such a supposition would be both inaccurate and irresponsible. Conversely, the absence of ignitable liquid identification in a sample collected from a suspected arson should not preclude the determination of an incendiary fire. There are a variety of plausible reasons (Table 6.4) for both positive and negative laboratory findings and thus analytical results should be only one of the many tools used in the investigative process.

Laboratory analysis of fire debris can be a valuable tool to the fire investigator. The development of increasingly more sensitive extraction and analysis processes can allow for detection and identification of very minute amounts of residual ignitable liquids in debris. These findings can help substantiate the determination of fire cause, however they must be evaluated carefully.

References

1 J. Speight (1991) *The Chemistry and Technology of Petroleum*, 2nd edn, Marcel Dekker, New York.
2 J. Gary and G. Handwerk (1994) *Petroleum Refining Technology and Economics*, 3rd edn, Marcel Dekker, New York.
3 American Society for Testing and Materials (2002) ASTM E 1618–01 test method for identification of ignitable liquid residues in extracts from fire debris samples by gas chromatography–mass spectrometry, in *Annual Book of ASTM Standards, 2002*.
4 ExxonMobil' Exxon Chemical's Norpar Functional Fluids, www.exxonmobil.com, 7-1-02.
5 Normal Paraffins, www.sasoltechdata.com, 7-1-02.
6 E. Stauffer (2001) Identification and characterization of interfering products in fire debris analysis, Master's thesis, Florida International University.
7 D. Tranthim-Fryer and J. DeHaan (1997) Canine accelerant detectors and problems with carpet pyrolysis products, *Science and Justice* 37: 39–46.
8 R. Martin Smith, (1982) Arson analysis by mass chromatography. *Analytical Chemistry* 54: 1399A–1409A.
9 J. Nowicki (1990) An accelerant classification scheme based on analysis by gas chromatography/mass spectrometry (GC-MS). *Journal of Forensic Sciences* 35(5): 1064–1086.
10 M. Gilbert (1998) The use of individual extracted ion profiles vs summed extracted ion profiles in fire debris analysis. *Journal of Forensic Sciences* 43(4): 871–876.
11 R. Newman and M. Gilbert (1998) *GC-MS Guide to Ignitable Liquids*, CRC Press, Boca Raton.
12 J. Nowicki (1993) Automated data analysis of fire debris samples using gas chromatography–mass spectrometry and macro programming. *Journal of Forensic Sciences* 38(6): 1354–1362.
13 American Society for Testing and Materials (2002) ASTM E 1387-01 test method for identification of ignitable liquid residues in extracts from fire debris samples by gas chromatography, in *Annual Book of ASTM Standards, 2002*.
14 B. Levin (1986) A summary of the NBS literature reviews on the chemical nature and toxicity of the pyrolysis and combustion products from seven plastics. US Dept of Commerce, National Bureau of Standards, Gaithersburg, MD pp. 1–31.
15 Trimpe (1991) Turpentine in arson analysis. *Journal of Forensic Sciences* 36(4): 1059–1073.
16 B. Chanson, E. Ertan, O. Dlelmont, E. DuPasquier and J. Martin (2000) Turpentine identification in fire debris analysis, Second European Academy of Forensic Science Meeting, Cracow, Poland.
17 *d*-Limonene Product Data Sheet (2001) Florida Chemical Company, Winter Haven, Florida.
18 Ignitable Liquid Reference Database (2002) National Center for Forensic Science, www.ncfs.ucf.edu

19 J. Lentini (1998) Differentiation of asphalt and smoke condensates from liquid petroleum distillates using GC/MS. *Journal of Forensic Sciences* 43(1): 97–113.

20 P. Loscalzo, P.R. DeForest and J. Chao (1980) A study to determine the limit of detectability of gasoline vapor from simulated arson residues. *Journal of Forensic Sciences* 25(1): 162–167.

21 D. Mann and W. Gresham (1990) Microbial degradation of gasoline in soil. *Journal of Forensic Sciences* 35(4): 913–923.

22 D. Chalmers, X. Yan, A. Cassita, R. Hrynchuck, and P. Sandercock (2001) Degradation of gasoline, barbecue starter fluid, and diesel fuel by microbial action in soil, *Canadian Society of Forensic Science Journal* 34(2): 49–62.

23 J. Lentini, J. Dolan, and C. Cherry (2000) The petroleum laced background. *Journal of Forensic Sciences* 45(5): 968–989.

24 IAAI Forensic Science Committee (1989) Glossary of terms related to chemical and instrumental analysis of fire debris. *Fire and Arson Investigator* 40(2): 27.

25 J. Lentini (2001) Persistence of floor coating solvents. *Journal of Forensic Sciences*, 46(6): 1470–1473.

7

Sources of interference in fire debris analysis

Eric Stauffer

Introduction

Apparition of gas chromatography

The first article demonstrating the application of gas chromatography (GC) to fire debris analysis was published by Lucas in 1960 [1]. This marked the beginning of modern fire debris analysis. Prior to that, ignitable liquid residues (ILR) were analysed using rather unreliable methods [2–6].

Question about identification of ILR

The apparition of GC in fire debris analysis did not initially bring any doubts regarding the identification of ILR. It was not until eight years later, in 1968, that the first article questioning the identification of recovered compounds as originating from an ignitable liquid was published [7]. Ettling and Adams raised the issue of interference from substrate materials when they stated: 'For example, it is known that some hydrocarbons may be produced by pyrolysis of wood.' However, they also observed that the residues from burned wood, paper, and textiles are distinguishable from ignitable liquids.

Since then, several other authors have performed and published research on this topic. In 1976, Clodfelter and Hueske published an article comparing pyrolysis products with different 'accelerants' such as gasoline, diesel fuel, kerosene or jet fuel [8]. They concluded that the chromatograms of burned substrates are readily distinguishable from the chromatograms of the ignitable liquids tested.

In 1978, Thomas warned fire investigators about the necessity of collecting control samples when submitting fire debris samples to the laboratory for analysis [9]. He stated '...not all hydrocarbon vapors come from flammable liquids. [...] These hydrocarbons may also be detected by the GC but should not be confused by the analyst with any of the more common flammable liquids.'

Apparition of mass spectrometry

The next important step toward the modern analysis of fire debris is the implementation of mass spectrometry (MS) after separation by GC. In 1982 and 1983 Smith presented gas chromatography–mass spectrometry (GC–MS) as a valuable tool in discriminating ILR from pyrolyzates [10,11]. He reported the presence of styrene and low-boiling alkylbenzenes as arising from pyrolysis of styrene-containing polymers.

In 1984, Howard and McKague reported a case in which charred carpet presented a pattern very similar to the one offered by gasoline, but concluded that it was due to the pyrolysis products (PyP) [12]. They used both gas chromatography–flame ionization detection (GC–FID) and GC–MS to identify several aromatic products.

The same year, Stone and Lomonte discussed false positives in fire debris analysis and related a case report made by Nowicki in which an 'accelerant' was produced by the burning of flooring [13,14]. They stated that often turpentine is detected in wood samples, which is due to PyP. They also related a case where a naphtha or diesel was found in fire debris, which again was produced by the pyrolysis of the substrate. They recommended GC–MS versus GC alone to decrease the number of false positives.

In 1988, DeHaan and Bonarius produced the first complete study on PyP [15]. They offered a valuable set of solutions to discriminate between PyP and ILR.

Bertsch presented an interesting study in 1994 describing PyP released by burned carpet and carpet padding [16]. He showed the presence of styrene, methylstyrene, small amounts of aromatics and high presence of naphthalenes and methylnaphthalenes. He used the extracted ion profiles to discriminate PyP from ignitable liquids based on the ratios of the different isomers.

Keto, in 1995, advocated the use of extracted ion chromatograms in order to help identify ILR in the presence of contaminated arson debris [17]. He cautioned, 'Another hazard is that petroleum like isomer profiles may not originate from petroleum distillates at all.'

In 1996, Kurz *et al.* tested accelerant canines' ability to discriminate between PyP and ILR [18]. They concluded that PyP generated by burning substrates contained common substances to the ones present in gasoline, which can distract the canines.

In 1997, Tranthim-Fryer and DeHaan presented an identification of PyP of carpet and underlay that resulted in false positive alerts for accelerant detection canines [19].

In 1998, Lentini demonstrated the difference between PyP from tar materials (asphalt) found in smoke condensates and heavy petroleum distillates [20].

In 2000, Lentini, Dolan, and Cherry presented a study on petroleum-laced background and demonstrate that petroleum-based products are inherently present in samples prior to burning [21].

In 2002, Cavanagh, Du Pasquier, and Lennard presented an evaluation of the background contamination of carpet subjected to exposure to external environment for a certain amount of time [22]. They advised the use of comparison samples.

The same year, Fernandes presented an article comparing burned and unburned substrates [23].

Misconception about 'pyrolysis products'

It is very often encountered throughout the literature and analytical reports that the main difficulty in fire debris analysis is the presence of PyP and that these products interfere with the proper identification of ignitable liquids. Pyrolysis products are usually blamed as the sole source of interferences in fire debris samples. This concept is wrong, or at least, it is potentially poor verbiage.

Pyrolysis products are only a portion of the products that are released by a burned substrate that will interfere with ILR recovery and identification. Very few authors have realized that phenomenon throughout their publications [16,21–24]. However, the distinction between the different sources has never been reliably demonstrated. Lentini, Dolan, and Cherry successfully demonstrated the existence of background products in different substrates as did Cavanagh, Du Pasquier, and Lennard [21,22].

This chapter covers the sources of interference in fire debris analysis. First, the concept of interfering products in fire debris analysis is introduced and explained. Second, the different sources of interfering products are presented one by one and thoroughly explained. Finally, some practical cases are presented and analysed. This should allow the reader to put into practice the concepts learned in this chapter as well as those in Chapters 5 and 6. The understanding and knowledge of the concept of interfering products will improve the analysts' skills.

Concept of interfering products

In the field of fire debris analysis, the term 'interfering products' is defined as 'the set of products found in a sample that interfere with the proper identification of ignitable liquids residues' [24]. It is possible to distinguish three categories of products originating from a substrate that will interfere with ILR recovery and analysis.

The three categories are

- Substrate background products (SBP)
- Pyrolysis products (PyP)
- Combustion products (CoP).

By applying the rules of a substrate's composition, manufacture, contamination, and combustion, it is possible to easily understand the different categories of interfering products released.

Figure 7.1 is a schematic representation of the sequence of events that a substrate undergoes during a fire accelerated with an ignitable liquid poured onto a substrate.

Pre-fire phase

Let's consider a substrate, such as a nylon carpet. This substrate contains SBP as explained later in the chapter. A liquid, such as gasoline, is poured onto the substrate. Through gravitational and capillary forces, part of the liquid will travel downward and into the deep layers of the substrate. It will get adsorbed onto the surface of the substrate and the excess will create a puddle of liquid. Also, with enough liquid and time, it will eventually reach the wood or concrete under the carpet. Part of the liquid will also evaporate at a rate depending on the physical properties of the liquid and the environmental conditions.

Ignition phase

Ignition occurs and the vapours above the liquid catch fire. Heat is released due to the combustion process. If the liquid has formed a puddle, the radiant heat will first affect the

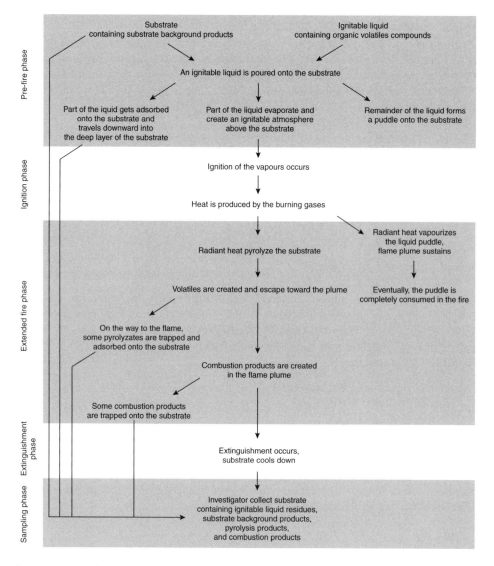

Figure 7.1 A schematic representation of the sequence of events during a fire accelerated with an ignitable liquid poured onto a substrate

evaporation of the liquid. This creation of vapours will feed the fire and eventually, the puddle will disappear.

Extended fire phase

The radiation will also heat the substrate onto which the liquid was poured. Since solids do not burn, they must first transform into gas. The substrate can use different phase transformation pathways, as shown in Figure 7.2. However, the most common pathway, and most pertinent in the scope of this chapter, is the pyrolysis of the substrate. While the

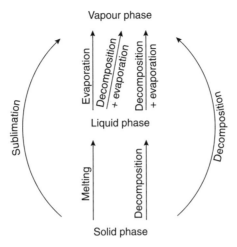

Figure 7.2 Phase transformation pathways

phenomenon of pyrolysis will be explained in detailed later in the text, it can be briefly stated that flammable volatiles are created by the thermal decomposition of the substrate. Due to buoyancy, they will rise to the plume of the flame and sustain the fire. On the way to the plume, it is suspected that some volatiles get trapped onto the carbonized substrate.

Extinguishment phase

At any time during the fire, extinguishment can happen. The fire can self-extinguish by running out of fuel or oxygen, or it can be put out by an external act. This can be accomplished by reducing the oxygen, decreasing the energy of the system, or separating the fuel from the oxygen and/or energy source [25].

Sampling phase

When the fire investigator works the fire scene, he/she will develop a hypothesis regarding the origin and cause of the fire. Eventually, in this case, a sample of the carpet will be collected and submitted to the forensic laboratory for fire debris analysis [26]. The sample collected contains ILR from the gasoline poured, as well as SBP, PyP, and CoP.

Different sources of interfering products

The different sources of interfering products are shown in Figure 7.3.

Substrate background products

This category includes all the products (usually petroleum based) that are already adsorbed onto the surface of the substrate prior to burning that will interfere in the analysis of fire debris. It is possible to distinguish three different sub-sources from which these products originate, as shown in Figure 7.3.

195

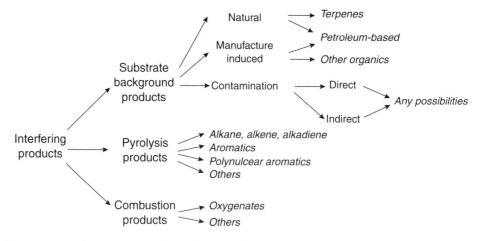

Figure 7.3 Different sources of interfering products

Natural

There can be some natural background present in the raw material of the substrate. As an example, some rubbers contain some high-boiling point hydrocarbons. Also, it is very well-known that some woods contain terpenes as part of their natural background [27,28].

Manufacture induced

The manufacturing and/or finishing process may also involve the use of products that will stay adsorbed onto the substrate and later interfere with ILR identification. Examples of such products are solvents used in the manufacturing process for different purposes [29]. Another known example is the use of kerosene in the printing industry as a solvent for the ink [21].

Contamination

It is very simple to imagine that contaminations of any substrates can occur naturally and accidentally every day. These contaminants can be further classified in two different groups:

Direct contact   when there is an established contact between two objects or two surfaces. This is known in forensic sciences as The Principle of Exchange from Locard: 'tantôt le malfaiteur a laissé sur les lieux les marques de son passage, tantôt, par une action inverse, il a emporté sur son corps ou sur ses vêtements les indices de son séjour ou de son geste' [30]. While this principle was expressed as describing the exchange of material and traces between the author of a crime and the crime scene, it can be very easily adapted to a regular contaminating situation. As an example, a shoe walking on gasoline traces at a gas station will leave contamination on the floor mat of the vehicle by direct contact. This has been successfully demonstrated by Cavanagh, Du Pasquier, and Lennard [22].

Indirect contact   when the contamination is brought to the substrate by aerial transportation. An example of such contamination is the use of positive-pressure ventilation systems at fire scenes. A study made by Lang and Dixon shows that it was possible that gasoline vapours could contaminate some substrates inside a house through the use of a positive pressure ventilation fan [31]. On the other hand, Koussiafes, using different conditions, did not observe that phenomenon [32].

Pyrolysis products

The second most important category of products is represented by the PyP released by the substrate when subjected to heat. Pyrolysis can be defined as 'a process, by which a solid (or a liquid) undergoes degradation of its chemicals into smaller volatile molecules under heat, without interaction with oxygen or any other oxidants, that is necessary for almost all solids (or liquids) to burn.' [24]

While pyrolysis is a process that produces many different chemicals and that is highly influenced by environmental conditions as well as by the nature of the substrate involved, it commonly follows three main degradation mechanisms; random scission, side-group scission, and monomer reversion [33,34]. Other peculiar mechanisms have also been identified, but are of less interest to the fire debris analysts.

Random scission

This first mechanism involves the random breaking of the polymer's backbone (C–C bond) [35]. Figure 7.4 shows the random scission of polyethylene. This scission results in the

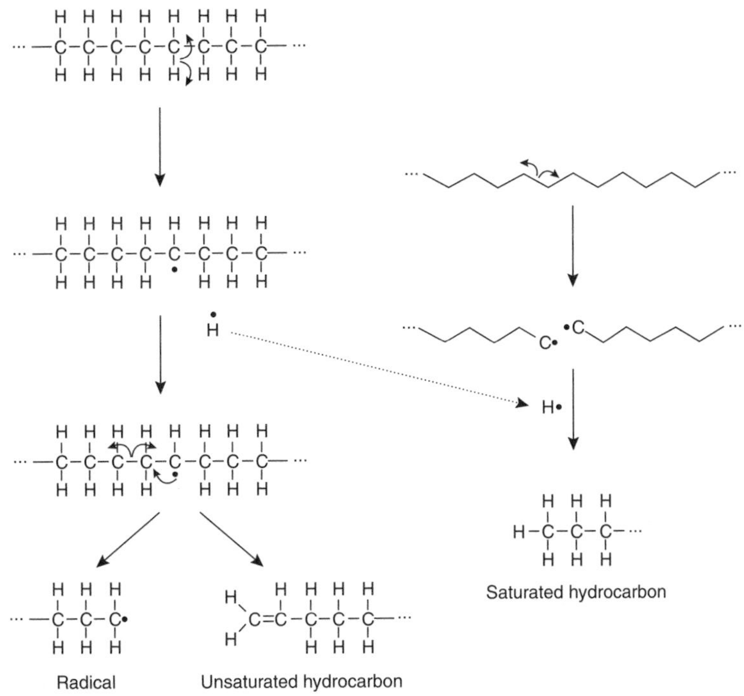

Figure 7.4 Random scission of polyethylene (reproduced with permission from Science & Justice, Forensic Science Society)

197

production of series of alkanes, alkenes, and alkadienes of different lengths. The chromatographic result of such PyP is shown in the previous chapter in Figure 6.24 with the example of polyethylene.

Side-group scission

This mechanism involves the breakage of side groups away from the backbone chain [36]. Hence, the backbone becomes polyunsaturated and undergoes a rearrangement into aromatics, as shown in Figure 7.5 with the example of polyvinylchloride.

The products from this type of pyrolysis include a wide variety of aromatics ranging from benzene to C4- or C5-alkylbenzenes, including an even greater variety of polyaromatic hydrocarbons.

Monomer reversion

This last common degradation mechanism is probably the least interesting to fire debris analysts. The polymer simply unzips and, therefore, reverts to its monomeric version, as shown in Figure 7.6 with the example of an acrylate polymer. This produces usually one compound that is very predictable when the polymer structure is known [37]. The monomeric reversion, itself, does not usually interfere too much with the identification of ILR, because it presents only one peak that is typically not common to an ignitable liquid.

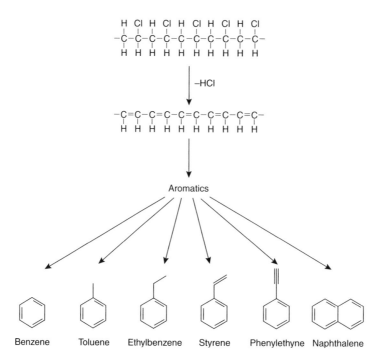

Figure 7.5 The mechanism of side-group scission and rearrangement into aromatics (reproduced with permission from Science & Justice, Forensic Science Society)

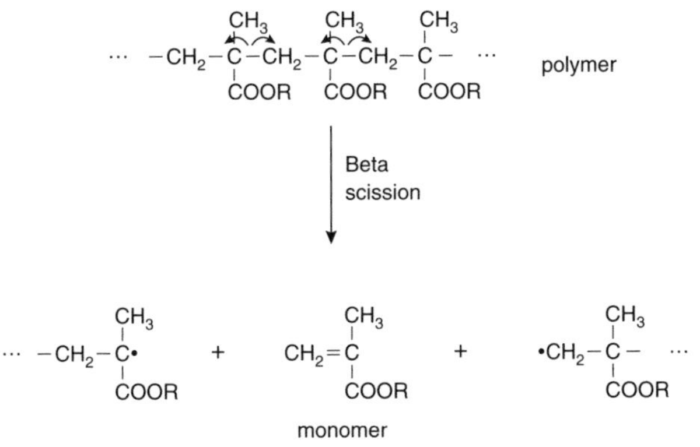

Figure 7.6 The reversion of polymeric to monomeric form (reproduced with permission from Science & Justice, Forensic Science Society)

Pyrolysis involving multiple degradation mechanisms simultaneously

As shown previously, degradation mechanisms are pretty straightforward and simple to apply. Pyrolysis products can be predicted and are well defined for a given polymer. However, nature being nature, it is much more complicated than that. As a matter of fact, very few polymers undergo pyrolysis using only one degradation mechanism. Commonly, two or more mechanisms will occur simultaneously.

So, how is it possible to determine which polymer undergoes which pyrolysis mechanisms? Simply put, it is by following the law of the weakest bond. The weakest bond in the polymer will usually break first and consequently determine the mechanism. Unfortunately, parameters, such as rate of temperature rise, will slightly change the bond breakage order in some instances. However, by looking at bond dissociation energy, it is possible to understand the products released by different polymers [38].

Figure 7.7 shows a schematic representation of different polymers placed according to their degradation mechanisms.

Figure 7.8 shows a list of some of the most commonly encountered polymers in a typical household environment. Each polymer is presented with its chemical structure, its main pyrolysis degradation mechanism(s), and the typical PyP released.

The careful study of this table should allow any scientist to get accustomed to the kind of polymers encountered in fire debris analysis. The knowledge of polymer structure will help to improve the understanding of the subsequent PyP, and should improve greatly the interpretative skills of the scientist.

Combustion products

Combustion is a very complex phenomenon and is often described as a chain reaction. Some authors have shown very complex mechanisms of reaction [25,39]. Thus, different products can be released during combustion according to the conditions in which it occurs. If the conditions are ideal, the combustion will be complete and CoP will be completely oxidized and reduced. With most organic polymers, the ultimate CoP produced will be CO_2 and H_2O. However, when the oxidizer is present by default, it will lead to other products that are not completely oxidized or reduced, referred to as products of incomplete combustion.

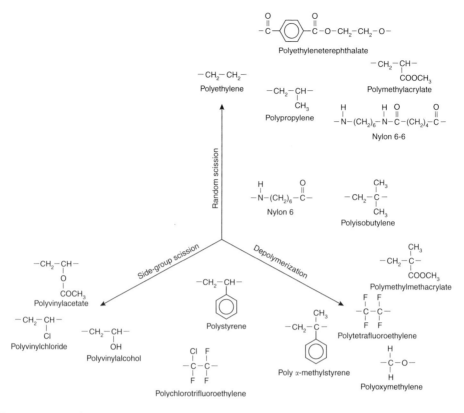

Figure 7.7 A schematic representation of polymers according to their degradation mechanisms (reproduced with permission from Science & Justice, Forensic Science Society)

The best example describing these incomplete CoP is shown with smoke condensates. They mainly contain polyaromatic hydrocarbons as described in the literature [40,41]. Differentiation between combustion and PyP is primarily based on the fact that PyP do not result from oxidation reactions, while CoP do. While this concept is disputable, particularly with the definition of some pyrolysis processes, it is beyond the scope of this chapter to adequately present the differences [42]. It is interesting to note that the recent revision of the ASTM standard E 1618 now integrates the wording 'pyrolysis and combustion products' in its section 11.2.1, while it presented only 'pyrolysis products' in the previous edition [43,44].

Knowledge of polymers and interfering products

Almost any substrate can contain SBP inherent to its composition. On the other hand, PyP are almost exclusively produced by polymers. Nevertheless, polymers constitute most of the raw material used in household items. Hence, it is important for the fire investigator and the fire debris analyst to know the different polymers found among fire debris.

The following is a list of the most commonly encountered polymers and their common applications [45,46]. Their structures are shown in Figure 7.8.

	Chemical structure	Degradation mechanism	Pyrolysis products
Polyolefins			
Polyethylene	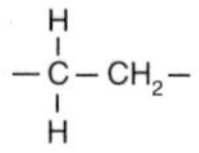 $-\overset{\text{H}}{\underset{\text{H}}{\text{C}}}-CH_2-$	random scission	n-alkanes n-alkenes n-alkadienes
Polypropylene	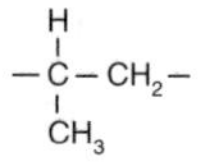 $-\overset{\text{H}}{\underset{CH_3}{\text{C}}}-CH_2-$	random scission	branched alkanes branched alkenes branched alkadienes
Polyisobutylene	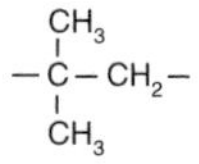 $-\overset{CH_3}{\underset{CH_3}{\text{C}}}-CH_2-$	random scission monomer reversion	branched alkanes branched alkenes branched alkadienes
Styrene polymers			
Polystyrene	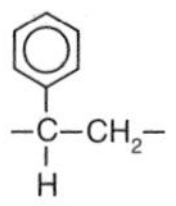 $-\overset{C_6H_5}{\underset{\text{H}}{\text{C}}}-CH_2-$	monomer reversion side-group scission	styrene aromatics
Poly α-methylstyrene	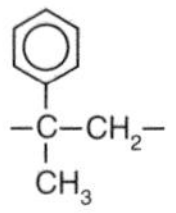 $-\overset{C_6H_5}{\underset{CH_3}{\text{C}}}-CH_2-$	monomer reversion	α-methylstyrene
Vinyl polymers			
Polyvinylchloride	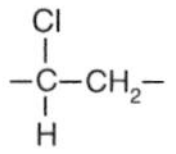 $-\overset{Cl}{\underset{\text{H}}{\text{C}}}-CH_2-$	side-group scission	aromatics chlorinated compounds
Polyvinylidenechloride	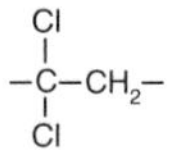 $-\overset{Cl}{\underset{Cl}{\text{C}}}-CH_2-$	side-group scission	aromatics chlorinated compounds
Polyvinylidenefluoride	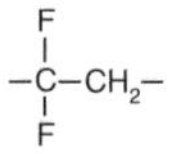 $-\overset{F}{\underset{F}{\text{C}}}-CH_2-$	side-group scission	aromatics fluorinated compounds
Polytetrafluoroethylene	$-CF_2-CF_2-$	monomer reversion	tetrafluoroethylene
Polyvinylacetate	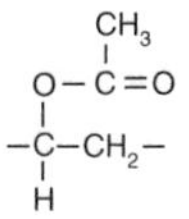 $-\overset{O-C(=O)CH_3}{\underset{\text{H}}{\text{C}}}-CH_2-$	side-group scission	aromatics acetic acid
Polyvinylalcohol	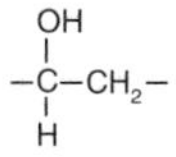 $-\overset{OH}{\underset{\text{H}}{\text{C}}}-CH_2-$	side-group scission	aromatics
Acrylate polymers			
Polymethylacrylate	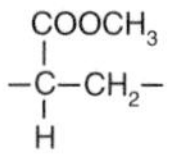 $-\overset{COOCH_3}{\underset{\text{H}}{\text{C}}}-CH_2-$	random scission	methanol oxygenated compounds
Polymethylmethacrylate	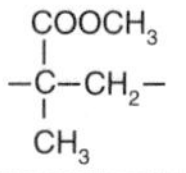 $-\overset{COOCH_3}{\underset{CH_3}{\text{C}}}-CH_2-$	monomer reversion	methylmethacrylate

Figure 7.8 A list of some common polymers, their chemical structure, degradation mechanisms, and pyrolysis products

	Chemical structure	Degradation mechanism	Pyrolysis products
Polyamides			
Nylon 6	$-(CH_2)_5-\overset{\overset{O}{\|}}{C}-\underset{\underset{H}{\|}}{N}-$	random scission accompanied by cross-linking	oxygenated compounds
Nylon 6-6	$-\overset{\overset{O}{\|}}{C}-(CH_2)_4-\overset{\overset{O}{\|}}{C}-\underset{\underset{H}{\|}}{N}-(CH_2)_6-\overset{\overset{H}{\|}}{N}-$	random scission accompanied by cross-linking	oxygenated compounds
Polyesters			
Polyethyleneterephthalate	$-\overset{\overset{O}{\|}}{C}-\langle\bigcirc\rangle-\overset{\overset{O}{\|}}{C}-O-CH_2-CH_2-O-$	random scission (of the CO bonds)	oxygenated compounds
Polyethylenaphthalate	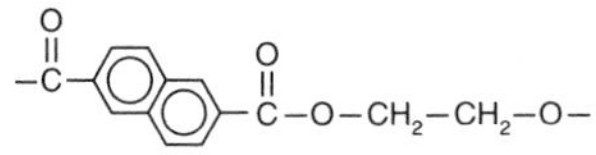 random scission (of the CO bonds)		oxygenated compounds
Polybutyleneterephthalate	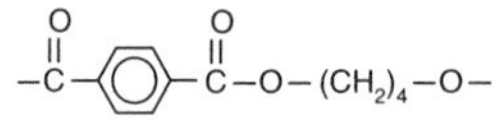 random scission (of the CO bonds)		oxygenated compounds
Polycarbonates			
	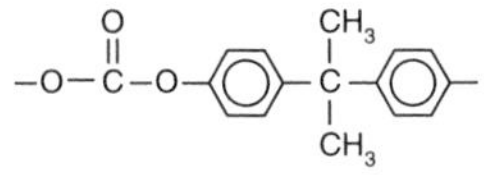 side-group scission		aromatics chlorinated compounds
Polyurethanes			
	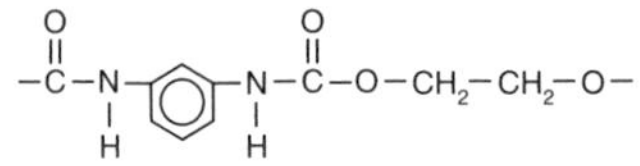 monomer reversion random scission		isocyanides oxygenated compounds
Rubbers			
Polybutadiene	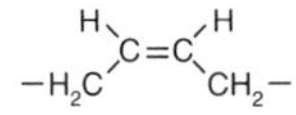	monomer reversion (followed by dimerization)	butadiene alkenes
Polyisoprene	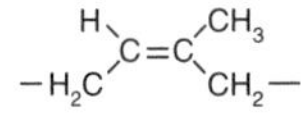	monomer reversion (followed by dimerization)	dipentene alkenes
SBS	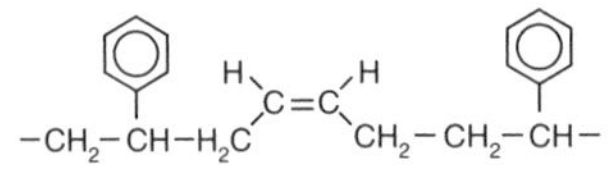	monomer reversion random scission	butadiene alkenes aromatics
Polychloroprene	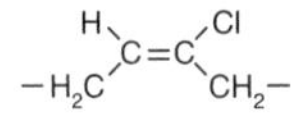	side-group scission	aromatics
Polyacrylonitriles and copolymers			
Polyacrylonitrile	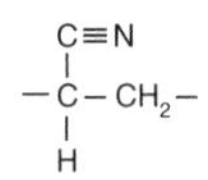	side-group scission	aromatics
SAN	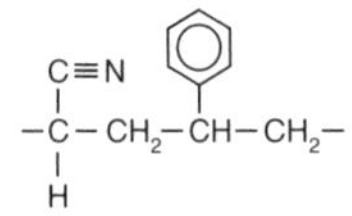	side-group scission random scission	aromatics
Cellulose			
	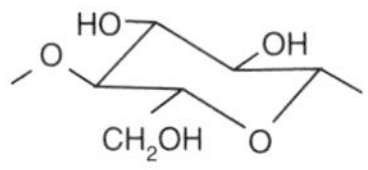	random scission of the CO bonds	oxygenated compounds

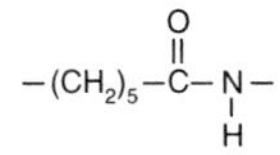

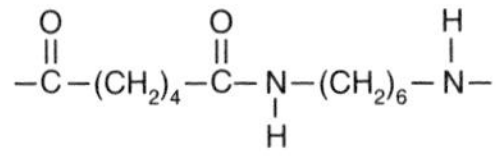

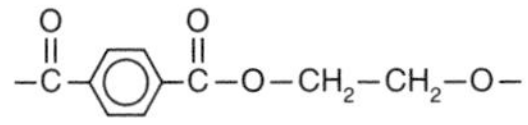

Figure 7.8 (Continued)

Polyolefins

Polyethylene (PE)

This is the simplest polymer. It is constituted by the repetition of an ethylene group. Sometimes, it is referred to low density (LDPE) or high density polyethylene (HDPE). This depends on the degree of branching between two chains. The more branching, the lower the density. HDPE is much stronger than LDPE, and is therefore used to make highly resistant fibres, sheets, or objects. For example, PE is used to manufacture grocery and rubbish bags, plastic boxes such as Rubbermaid®, buckets, hoses, shampoo bottles, children's toys, car gas tanks, and even some small backyard shelters. It is also used for the insulation of electrical wiring.

Polypropylene (PP)

When a methyl group is added to one of the carbons of the repetitive unit in PE, PP is obtained. PP is a very versatile polymer, used both as a plastic and as a fibre. Typical objects made of PP are dish-washer resistant food containers, some plastic parts in vehicles (garnitures, HVAC duct, etc.), one-time use objects, and plastic parts in electronic apparatuses. Also, as a fibre it is used to manufacture indoor–outdoor carpets. Most of the outdoor carpets are made of PP, since it does not absorb water.

Polyisobutylene (PIB)

When two methyl groups are attached to one of the carbons in PE, PIB is obtained. This is an elastomer that is also often classified as a rubber. The main quality of PIB is the fact that it is airtight. This is why it is used for the inner liner of tires and basketballs.

Styrene polymers

Polystyrene (PS)

The repetitive element is styrene, which is obtained by substituting a phenyl group for a hydrogen in the repetitive element of PE. This is the second most common plastic among the household items. It is used to make computer housings, model cars, plastic parts of all kinds of appliances, one-time use cutleries, toys, CD cases, keys on a keyboard, and moulded parts inside a vehicle. Perhaps, more widely known is polystyrene foam, which is used with great versatility. Examples of such use are drinking cups, packing material for prrotection during shipping, and building material.

Vinyl polymers

Polyvinylchloride (PVC)

When a chlorine atom replaces a hydrogen in the repetitive element of PE, PVC is obtained. PVC is used nearly everywhere in the household. PVC is widely used to make objects that need to resist to water. Water pipes, vinyl siding, vinyl car tops, linoleum floor, raincoats, and shower curtains are some examples of the use of PVC. It is also very resistant to fire, due to the fact that when it pyrolyses, the chlorine atoms released inhibit combustion. Thus, PVC burns very badly and can barely sustain a flame by itself.

Polyvinylidenechloride (PVDC)

By replacing both hydrogens attached to the same carbon in PE with chlorine atoms, polyvinylidenechloride is obtained. PVDC is the plastic wrap that is used to wrap food. A brand name of PVDC is Saran®.

Polyvinylidenefluoride (PVDF)

When the two hydrogens from one of the carbon in PE are replaced with two fluorine atoms, it becomes polyvinylidenefluoride. PVDF offers excellent resistance to heat and electricity, which makes it a good material to insulate electrical wires. It is mainly used on wires that get hot, such as the ones inside a computer or other electronic apparatuses, when PE is not a suitable material. It is also used on wires in airplanes because it is fireproof. Since it offers good UV radiation resistance, it is blended with PMMA (see later) to make outside windows. Its chemical resistance is also excellent, which makes it a material of choice for bottles or containers to store chemicals.

Polytetrafluoroethylene (PTFE)

When the four hydrogens of PE are replaced with four fluorine atoms, it becomes PTFE, also known by the brand name Teflon®. Since it has the property of not sticking to anything, it is used to make nonstick cooking pans. It is also used to treat carpets and fabrics (such as Gore-Tex®) to make them stain resistant. It is also a well-accepted substance by the human body, and this is why some artificial body parts are made of PTFE.

Polyvinylacetate (PVAc)

When one of the hydrogen atoms in PE is replaced by an acetate group, it becomes PVAc. PVAc is well spread throughout the household items, however, it is not very well known. It is used to make adhesives, such as wood glues. Also, paper and textiles often have coatings made of PVAc, which make them shiny. PVAc is also one of the constituents forming the latex in acrylic latex paints.

Polyvinylalcohol (PVA)

By replacing a hydrogen atom in the repetitive element of PE by an alcohol group, it becomes PVA. It is used with polyethyleneterephthalate (PET) to make bottles for carbonated beverages. Carbon dioxide can easily go through PET, but not through PVA. By creating a sheet of layers of PET and PVA, a strong bottle that retains carbon dioxide is obtained.

Acrylate polymers

Polymethylmethacrylate (PMMA)

PMMA has a slightly more complicated structure and is shown in Figure 7.8. It is a clear plastic. Its main use is known as Plexiglass®, which replaces glass as it is shatterproof. It is used in a great variety of applications, such as hurricane-proof windows, and vehicle headlights. It is also used to make surfaces of hot tubs, shower units, and sinks. Indeed, counter tops that need to be resistant to heat are made of PMMA, usually mixed with an aluminium oxide. This is not to be confused with Formica®, which is a melamin-formaldehyde resin. Latex paints contain PMMA. Finally, it is used as an additive to oils and lubricants, since the addition of PMMA decreases the viscosity of these fluids.

Polyamides

Nylon 6

The structure of Nylon is shown in Figure 7.8. Nylon is probably the most common polymer used as a fibre. It is mainly used to make clothing and carpets. However, it is also used to make ropes, straps, bags, and parachutes.

Nylon 6-6

While there is a structural difference between Nylon 6 and Nylon 6-6, there is almost no difference in the physical properties. Nylon 6-6 was first invented and patented by DuPont™, so other manufacturers had to invent Nylon 6.

Polyesters

Polyethyleneterephthalate (PET)

PET is a repetitive element made of a terephthalate and an ethylene group, as shown in Figure 7.8. One very common application of this polymer is the shatterproof plastic bottles for beverages (used with PVA for carbonated drinks). It is also used to make some plastic jars. Usually items made of PET are not reusable, simply because the sterilization procedures involve temperatures too high for PET. It would melt and become too soft to keep its shape. PET is also used as a fibre to make some clothing.

Polyethylenenaphthalate (PEN)

PEN has a similar structure to PET, with a naphthalate group in lieu of the terephthalate group. PEN has a much higher glass transition temperature, which means it is much more resistant to high temperature than PET. This is why it is used in mixture with PET to create reusable bottles and jars.

Polybutyleneterephthalate (PBT)

PBT has a similar structure to PET with a butylene group in lieu of the ethylene group. It is used in the same kinds of applications as PET.

Polycarbonates

There are a couple of important polycarbonates, the most important of which is polycarbonate of bisphenol A. Its structure is shown in Figure 7.8. It is a clear plastic used to make shatterproof windows, lightweight eyeglass lenses, and some vehicle headlights.

Polyurethanes

There are several different polyurethanes, and the general structure is shown in Figure 7.8. These polymers are the most versatile and can be found as foam, fibres, paints, elastomers, or adhesives. A well-known polyurethane is Spandex®. It is used to make fabric that stretches, such as sport clothes. Polyurethane polymers are also used to manufacture the foam inside upholstered furniture such as couches and padded chairs, and other foams such as carpet padding, synthetic sponges, and material used for the soft insides of stuffed animals.

Rubbers

Polybutadiene (PBD)

PBD is formed of a monomer with two carbon-to-carbon double bonds. This polymer is one of the first synthetic elastomers developed. It is used as a rubber in a great variety of parts, because it resists cold temperatures better than some other polymers. It is used to manufacture hoses, belts, gaskets, and other car parts. It is also used as a copolymer in order to make tire tread. Among its uses as a copolymer, the most important is poly(styrene–butadiene–styrene), also known as SBS (see later).

Polyisoprene (PIP)

PIP has a slightly more complicated repetitive element as shown in Figure 7.8. As a matter of fact, PIP is natural rubber. It was originally extracted from the Hevea tree. It is used to manufacture tires following vulcanization. Objects commonly known to contain rubber, such as boots, cushioning, or rubber parts in a vehicle, have polyisoprene. Also, the sides of a tire are usually made out of PIP. It is also used to isolate electrical wiring and to make flexible elastic bandages, or rubber bands.

Poly(styrene–butadiene–styrene) (SBS)

SBS is a block copolymer that is made of a mixture of polystyrene and polybutadiene. It is also called hard rubber, since it has the rubbery aspect from PBD, but has been hardened due to the addition of PS. It is typically used for tire treads, shoe soles, or other items where durability is important.

Polychloroprene (PC)

By replacing one of the hydrogens attached to a central carbon in PIP with a chlorine, polychloroprene is obtained. It is better known as Neoprene® on the market. It is a rubber or elastomer and it is very resistant to oil. It is typically used to make wet suits for scuba diving.

Polyacrylonitriles and copolymers

Polyacrylonitrile (PAN)

When a nitrile group is substituted for a hydrogen in the repetitive unit of PE, polyacrylonitrile is obtained. It is not widely used alone. Although, it is used as a precursor to make carbon fibres, it is primarily used as a copolymer as shown next.

Poly(acrylonitrile-co-vinylchloride)

When PAN is mixed with PVC, a flame-retardant polymer is obtained. It is often used as a fibre, usually called modacrylic.

Poly(acrylonitrile-co-methylacrylate) and
Poly(acrylonitrile-co-methylmethacrylate)

When PAN is mixed with PMA or PMMA, another polymer is obtained. This is used mostly as a fibre for acrylic clothing.

Poly(styrene-co-acrylonitrile) (SAN)

This is a simple copolymer of PS and PAN. It is used as a plastic.

Poly(acrylonitrile-co-butadiene-co-styrene) (ABS)

This copolymer presents a much more complicated structure. As a matter of fact, it is a polybutadiene chain with SAN chains grafted onto it. It is used to manufacture moulded parts such as the bumpers of a car. It is a very light-weight plastic.

Cellulose

Cellulose is a natural polymer made of repeating units of glucose, as shown in Figure 7.8. It is the constituent of wood, paper, and cotton. It is used in an enormous variety of applications, such as clothing, books, furniture, etc.

Interfering products found in household items and their prediction

It is possible to partially predict interfering products according to the substrate studied. These predictions are, of course, to be taken with great caution. This exercise requires a lot of practical training that cannot be obtained through the reading of a book chapter. It is a self-taught experience that can be done very easily at the laboratory. The best way to do this is to obtain samples representative of fire debris samples that do not contain any ignitable liquids. This can be achieved by asking the fire investigator to bring back samples from fire scenes that are known not to contain ILR, or to obtain unburned samples and burn them at the laboratory.

If it gets very difficult to predict interfering products in some instances, it should be possible to understand and justify their presence. Again, the experience of looking at negative samples will allow the scientist to get accustomed to recognizing the interfering products at the earliest stage.

Prediction of substrate background products

While it is very difficult to predict contamination (subclass of SBP), the analyst will likely be able to find out what kind of natural and manufacture induced background are present. This was shown in some examples previously described. Understanding the concept of how the object or material is manufactured should allow the analyst to have an idea about the kind of products involved.

Prediction of pyrolysis products

All these polymers will be very likely to undergo combustion in a regular house fire, since the highest ignition temperature is around 500°C.

Pyrolysis products are a category that is pretty easy to predict, at least in a qualitative manner. However, there are many parameters that influence the creation of pyrolysis products.

Figure 7.8 also presents the different mechanisms of pyrolysis according to the polymer. Also, the second column shows the typical pyrolysis products encountered. By knowing the chemical structures of the polymers present in the fire debris analysed and by understanding which pyrolysis mechanisms applies to these particular polymers, it is possible to predict some of the resulting products. For a more detailed explanation of this approach, the reader is referred to another article written by the author [38].

Also, the scientist should be aware of the great variety of different polymers that can constitute a single object. For example, when referring to a nylon carpet, one should be mindful that the carpet fibres are made of nylon, but that the backing is probably made of polybutadiene and another natural fibre made of cellulose. Also, the carpet probably lies on a carpet pad made of polyurethane. Thus, the fire debris suddenly contains four different polymers, providing four different sets of pyrolysis products.

Prediction of combustion products

The literature does not contain any studies that address the exact differences between pyrolysis and combustion products in fire debris analysis. However, it seems that this category is relatively restricted and that most of the combustion products are oxidized version of pyrolysis products. Hence, these products are usually not similar to those found in ILR. Therefore, this category is not of much concern to the fire debris analyst.

Differentiation between SBP and PyP/CoP

In order to demonstrate the difference between SBP and PyP/CoP, identical unburned and burned substrates were extracted and analysed. Thus, it is possible to superimpose the two chromatograms obtained and isolate SBP from pyrolysis and combustion products. The three following chromatograms are examples of such results. It is advised to perform this kind of analysis with some of the most common items such as carpets, wood, and upholstery. This will help the analyst to quickly recognize the sources of the different products.

Polyester carpet

In Figure 7.9, it is possible to observe the difference between the chromatograms of unburned (blue line) and burned (red line) polyester carpet. Methyl isobutyl ketone (2.6 min), a C9 branched alkene (4.1 min), ethylbenzene (4.5 min), styrene (5.2 min), benzaldehyde (6.9 min), 2-ethyl-1-hexanol (9.3 min), naphthalene (13.8 min), cyclopentylbenzene (14.8 min) and some C15 branched alkenes (16–17 min) are produced by the pyrolysis and combustion

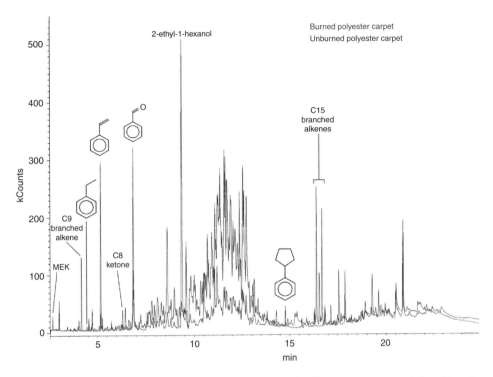

Figure 7.9 The chromatograms of burned and unburned polyester carpet (see Colour Plate I)

of the carpet. In this instance, toluene (3.0 min) is present in higher concentration in the unburned than in the burned sample. On the contrary, benzaldehyde is present in higher concentration in the burned sample than in the unburned one. Finally, it is possible to observe the same stack of aliphatics between 8 and 15 min (C10 to C13) in both chromatograms, showing that they are SBP.

Newspaper

A newspaper was extracted unburned (blue line) and burned (red line) as shown in Figure 7.10. Furaldehyde (3.9 min), ethylbenzene (4.5 min), phenylethyne (4.8 min), styrene (5.1 min), α-methylstyrene (7.6 min), benzofuran (8.0 min), indene (9.8 min), methoxyphenol (11.3 min), naphthalene (13.8 min) and methoxymethylphenol (14.1 min) are produced by the burning process of the newspaper. This is very consistent with the degradation process of cellulose. The heavy petroleum distillate pattern recovered toward the end of the chromatogram comes from the substrate background and has already been described in the literature by Lentini, Dolan, and Cherry [21].

Wood

An example of substrate that does not present a difference between the unburned (blue line) and the burned (red line) stage is yellow pine wood as shown in Figure 7.11. No extraneous peaks were produced in this instance with the burning of the substrate. The two chromatograms are very similar. As expected, most of the peaks are terpenes.

209

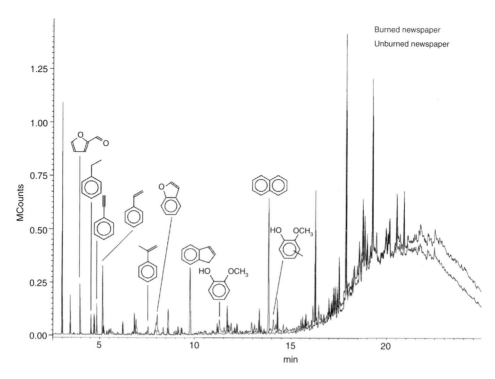

Figure 7.10 The chromatograms of burned and unburned newspaper (see Colour Plate II)

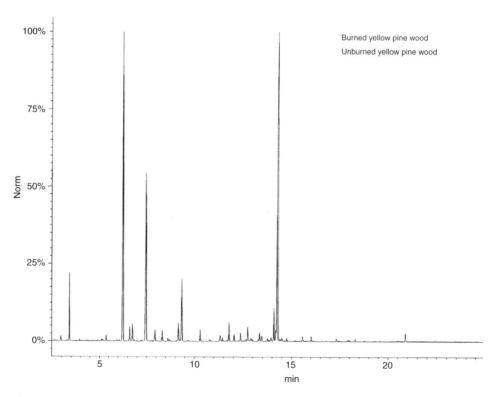

Figure 7.11 The chromatograms of burned and unburned yellow pine wood (see Colour Plate III)

Practical examples of interpretation of chromatograms and identification of sources of interferences

The following chromatograms were obtained from real samples that were analysed by the author. The data collected were interpreted using the information delivered in Chapters 6, and 7 of this book.

The purpose of this exercise is to train the investigator/analyst to look beyond the regular pattern of the possible ILR and to try to understand the sources of the different interfering products. By knowing the nature and the origin of interfering products, the interpretation of the chromatogram is simplified and, more important, the conclusion reached is much more certain.

All samples were extracted using passive headspace concentration according to ASTM standard E 1412-00 [47] at 90°C for 12–16 h on activated charcoal strip (ACS). The ACS was then desorbed using 0.5 mL of diethyl ether spiked at 100 ppm with tetrachloroethylene (PCE) as an internal standard. The extract was analysed on a Hewlett-Packard 6890-5973 GC–MS.

Before presenting the five cases, the reader is invited to look at Figures 7.12–7.15. Figure 7.12 shows the total ion chromatogram (TIC) of 75% evaporated gasoline, and the extracted ion profiles for the aromatic content are shown in Figure 7.13. The TIC of a heavy petroleum distillate ranging from C9 to C15 is shown in Figure 7.14 and the extracted ion profiles for both its aliphatic and aromatic contents are shown in Figure 7.15. These chromatograms can be compared to the case chromatograms for reference.

In Figure 7.13, the different ions represent the following groups of molecules: ion 105 for C2- and greater alkylbenzenes, ion 119 for C3- and greater alkylbenzenes, ion 117 for indan, ion 131 for substituted indans, ion 142 for methylnaphthalenes, and ion 156 for dimethyl- and ethylnaphthalenes.

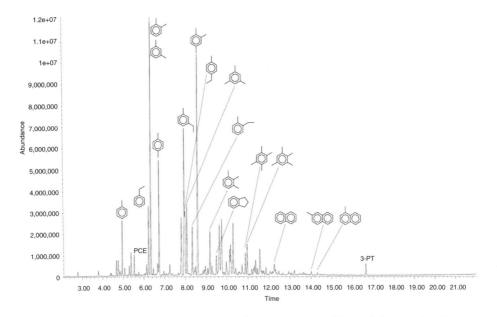

Figure 7.12 The TIC of 75% evaporated gasoline (PCE is tetrachloroethylene and 3-PT is 3-phenyltoluene, both internal standards)

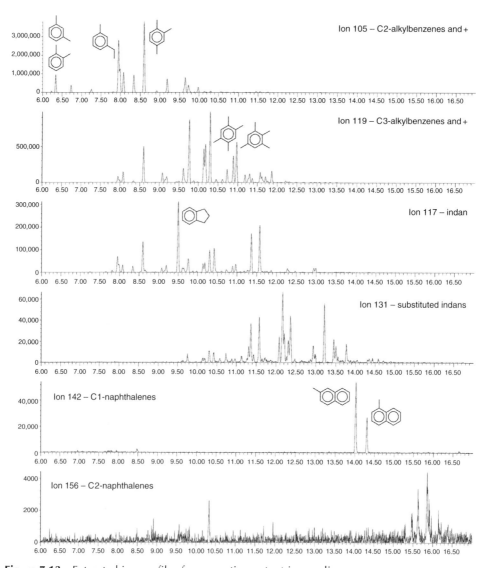

Figure 7.13 Extracted ion profiles for aromatic content in gasoline

In Figure 7.15, the different ions represent the following groups of molecules: ion 55 for alkenes, ion 57 for alkanes, ion 83 for cycloalkanes, and ion 105 for the aromatic compounds.

Case 1

This sample is a piece of carpet of unknown composition, collected from the floor of the room of origin of a house fire. Figure 7.16 shows the TIC of the extract.

At first sight, there is no obvious ignitable liquid pattern. It is possible to distinguish a few peaks between 5 and 9 min, followed by a little 'hump' between 10 and 13 min, a group of three peaks around 14 min, and finally a big peak around 19.5 min.

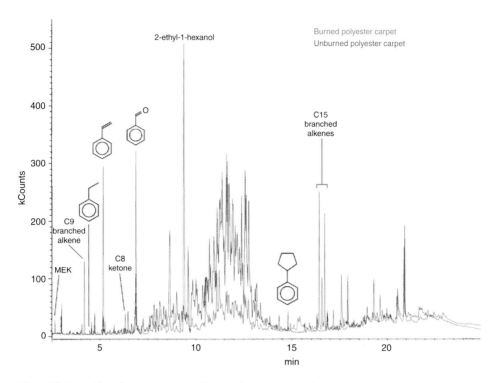

Colour Plate I The chromatograms of burned and unburned polyester carpet
(see Figure 7.9, p. 209)

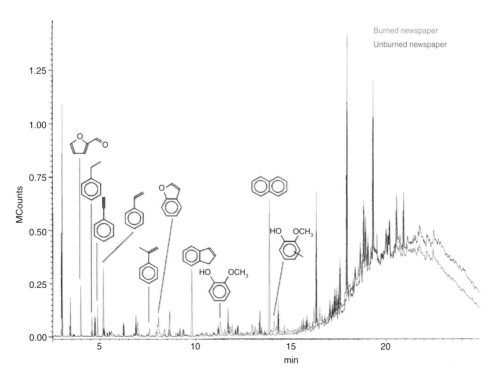

Colour Plate II The chromatograms of burned and unburned newspaper
(see Figure 7.10, p. 210)

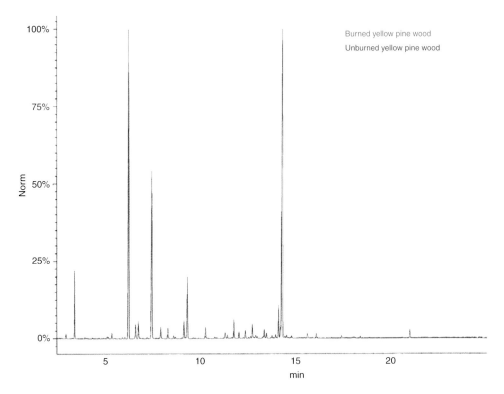

Colour Plate III The chromatograms of burned and unburned yellow pine wood
(see Figure 7.11, p. 210)

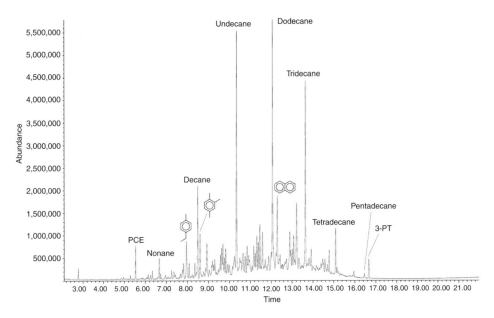

Figure 7.14 The TIC of a heavy petroleum distillate ranging from C9 to C15

The set of three peaks around 14 min is typical of carpet and are branched alkenes. They are found in a lot of different carpet samples, as shown previously in Figure 7.9.

In some instances, the little 'hump' can be confusing and might lead the analyst to think that a medium petroleum distillate is present. However, very often, the absence of a distinctive pattern of straight alkanes allows the rapid elimination of the presence of an MPD. This can be easily shown by extracting ions 55, 57, and 83, as presented in Figure 7.17. Furthermore, when looking at the individual peaks and identifying them using the mass spectral library, it is possible to quickly realize that most of these peaks are unsaturated and, therefore, unlikely to come from an ignitable liquid. As shown previously in the chapter, the hydrocarbon 'hump' around C10–C15 is often found as SBP in carpet sample.

Hence, the time span from 10 to 14 min was attributed to interfering products. The early eluting peaks in the chromatogram can be rapidly identified through a library search; 2,4-dimethyl-1-heptene (5.9 min), styrene (6.7 min), benzaldehyde (8.0 min), α-methylstyrene (8.3 min), and acetophenone (10.0 min) are all pyrolysis and combustion products. They do not originate from an ignitable liquid. Even 1,2,3-trimethylbenzene (8.6 min) can be produced by pyrolysis.

The final step consists in extracting ion profiles for aromatics. In this case, it did not reveal the presence of any ILR.

Case 2

This sample comes from a house fire. The fire was deliberately set and showed three separate areas of origin. Use of ignitable liquids was suspected due to the burn patterns. This is one of the samples collected by the author and analysed at the laboratory. It encompasses the remains of a trailer of unknown material that was used to spread the fire throughout the attic.

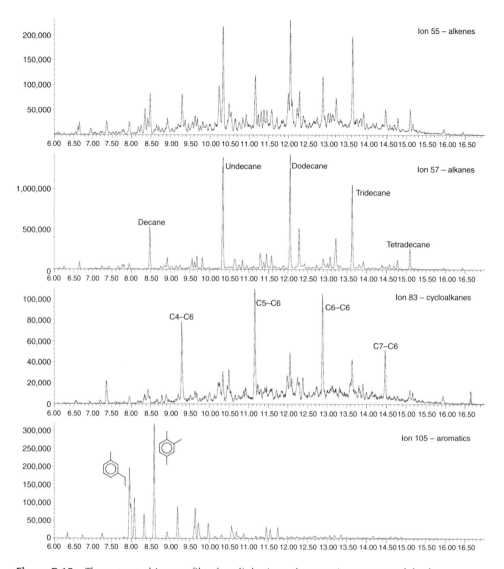

Figure 7.15 The extracted ion profiles for aliphatic and aromatic contents of the heavy petroleum distillate

Figure 7.18 shows the TIC of the extract. At first sight, it is possible to distinguish two patterns; an early-eluting one between 4 and 8 min, and a nice bell-shape pattern, of less intensity, between 8 and 18 min.

When the ions 55, 57, 83, and 105 are extracted, as shown in Figure 7.19, the pattern is immediately clearer. Again, without a close examination of the bell-shape pattern between 8 and 18 min, one would tend to conclude that there is a heavy petroleum distillate. However, a closer look at the pattern shows that there are doublets of peaks in the 55-window, which is indicative of pyrolysis products or polyethylene or asphalt. In this instance, the presence

214

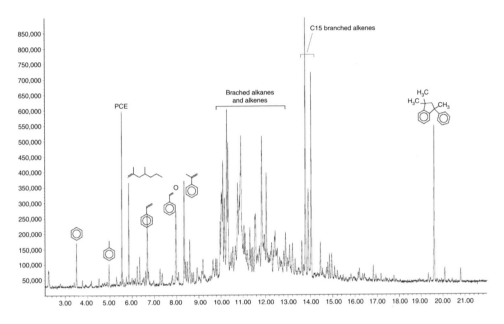

Figure 7.16 The TIC of the extract of a carpet sample obtained from a house fire (Case 1)

of doublets of peaks instead of triplets would suggest that it would originate from asphalt rather than polyethylene [20]. Also, the cycloalkanes in the 83-window do not originate from the pyrolysis of polyethylene [38]. All these elements suggest that the series of peaks from 8 to 18 minutes do not originate from an ignitable liquid.

When a closer look is taken at the early-eluting pattern, it is possible to distinguish a strong signal in the 57-window, which is almost perfectly mirrored in the 55-window. A library search of the different peaks in that region shows the presence of the following compounds: 2-methylhexane (3.5 min), 3-methylhexane (3.6 min), *n*-heptane (4.0 min), 2-methylheptane (4.9 min), 3-methylheptane (5.0 min), *n*-octane (5.3 min), 2,6-dimethylheptane (5.7 min), 2,3-dimethylheptane (6.1 min), 3-methyloctane (6.3 min), and *n*-nonane (6.7 min).

These products are not generated by the pyrolysis of polymers, and are not found in such quantity in substrate background. The presence of a light petroleum distillate ranging from C7 to C9 is, therefore, established.

In this instance, the study of the extracted ion profiles for the aromatic products did not reveal any pertinent data.

Case 3

This sample comes from the front driver's floorboard of a burned American car. It is extremely common for fire investigators to sample the floorboard of a vehicle when arson is suspected. The reasons behind that is the fact that most arsonists will basically spill an ignitable liquid throughout the seats of the vehicle and make a trailer on the floorboard up to the door, where they can light it in a safely manner. When the vehicle sustains the subsequent fire, all the upholstery and filling materials of the seats will often be completely consumed, and, usually, the only remaining debris is on the floorboard.

215

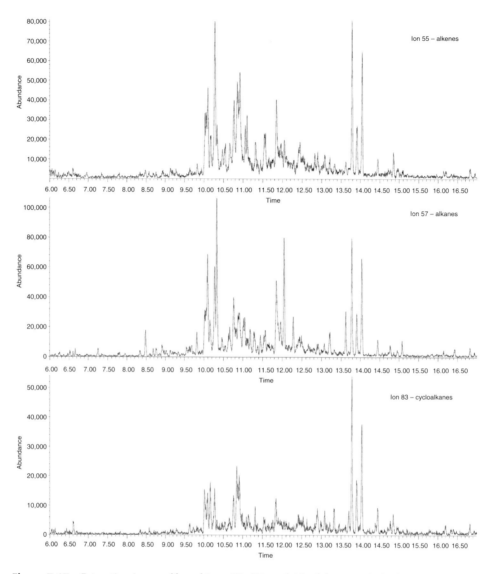

Figure 7.17 Extracting ion profiles of ions 55, 57, and 83 of the sample in Case 1

If the fire was not intense or long enough to completely burn out the floorboard, this kind of sample will usually contain a lot of interfering products. This is due to the fact that it encompasses the carpet as well as any objects that fell on the floor at some point in the fire. So, in considering the composition of the interfering products, one must take into account not only the complex composition of the carpet, but also the composition of fallen objects, which could include ABS, polystyrene and many other polymers. Therefore, these kinds of samples could include a wide collection of aromatic compounds led by the high presence of styrene and α-methylstyrene.

Figure 7.20 shows the TIC of the sample's extract. There are very few peaks and there is no obvious pattern of ignitable liquid.

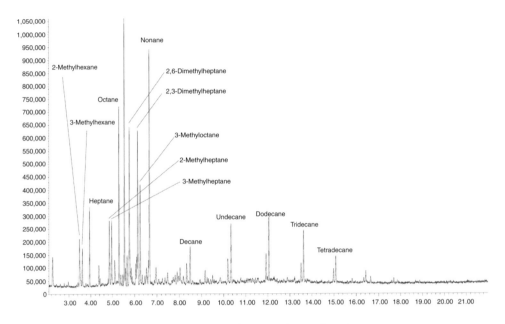

Figure 7.18 The TIC of the extract of sample from a house fire (Case 2)

It is possible to distinguish toluene (5.0 min), ethylbenzene (6.2 min), styrene (6.7 min), α-methylstyrene (8.4 min), and naphthalene (12.3 min). Indeed, there are other compounds such as methylheptene (5.9 min), phenol (8.1 min), and tetradecanedienol (10.2 min). All these compounds are interfering products originating from the different polymers constituting the passenger compartment of the vehicle.

Even when no obvious pattern is present, it is necessary to extract the ion profiles and confirm the presence or support the absence of ILR. Figure 7.21 shows the extracted ion profile for the ions 105, 119, 117, 131, 142, and 156.

In the 105 window, it is possible to notice that most of the C2-alkylbenzenes are present. Even if the styrene peak slightly masks the pattern, the ratios are not similar to the ones presented by the gasoline profile shown in Figure 7.13.

In the 119 window, it is possible to distinguish the weak signal given by most of the C3-alkylbenzene, but again in the wrong ratios.

The 117 and 131 windows, while mostly hidden by the tall α-methylstyrene peak, present very few indans and methylindans, common to the aromatic content of petroleum distillates. By taking a particular look at the 131 window, it is possible to realize that the peak ratios are not at all consistent with those of common petroleum products. When referred to Figure 7.13 it is possible to see that the proportions in between ions are not appropriate either.

The 142 window, representing the methylnaphthalenes, is very useful. In all petroleum distillates, the 2-methylnaphthalene (14.0 min) is greater than the 1-methylnaphthalene (14.3 min). If this ratio is inverted, it means that there is a contribution from the pyrolysis of the substrate. In this case they are equally great and indicate that they might originate from the pyrolysis process rather than an ILR. However, it is important to note that an inversion of this ratio does not mean that an ILR is not present; the methylnaphthalenes from an ILR could be present in addition to those produced from the substrate resulting in a skewed proportion.

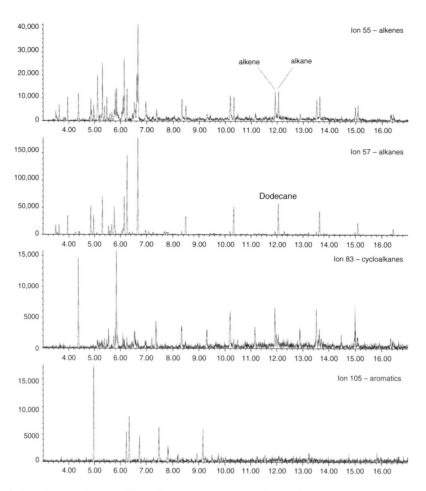

Figure 7.19 Extracted ion profiles of ions 55, 57, 83 and 105 of the sample in Case 2

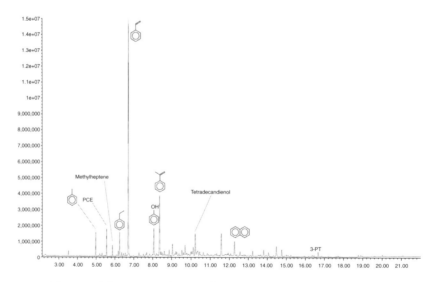

Figure 7.20 The TIC of the extract of a sample from a burned American car (Case 3)

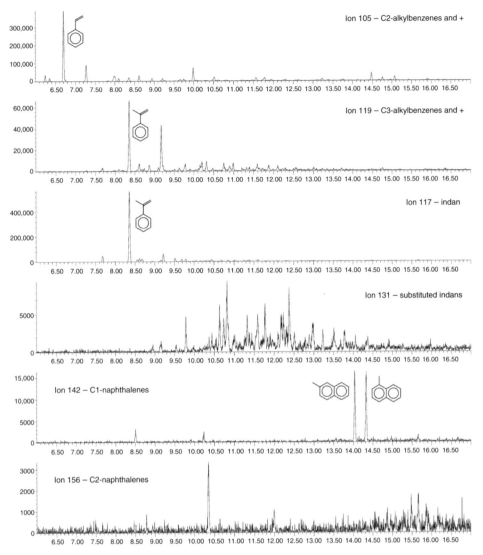

Figure 7.21 Extracted ion profiles for the ions 105, 119, 117, 131, 142, and 156 of the sample in Case 3

Finally, it is barely possible to distinguish any of the dimethyl- and ethylnaphthalenes in the 156 window. The extracted ion profile of ions 55, 57, and 83 did not reveal any ILR pattern. Therefore, it is possible to conclude that this sample failed to reveal the presence of any ILR. It solely contains interfering products typical of the polymers found in the passenger's compartment of a car.

Case 4

This sample consists of a piece of burned carpet removed from the room of origin of a house fire.

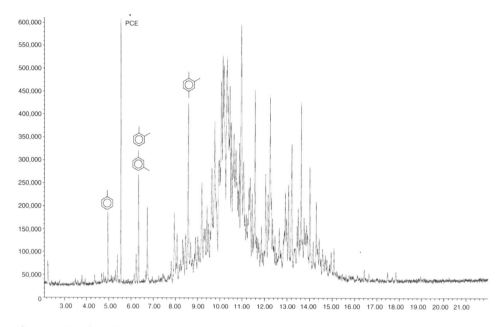

Figure 7.22 The TIC of the extract from a sample of burned carpet from a house fire (Case 4)

Figure 7.22 shows the TIC of the extract. At first sight, it is possible to distinguish a few peaks between 5 and 7 min, followed by a big "hump" between 8 and 16 min. While this hump is wider than the one presented in Figure 7.16, it is totally expected in a carpet, and is very likely made of branched alkanes and alkenes.

When the extracted ion profiles of ions 55, 57, and 83 are studied, it is possible to notice two distinct patterns, as shown in Figure 7.23. The first one is between 8 and 11.5 min and is represented by the typical hump found as SBP in most carpet. The 55 and 83 windows are very similar and the 57 window is weak and does not present any logical patterns.

By contrast there is an evident pattern of straight alkanes starting at 12.1 min with *n*-dodecane, followed by *n*-tridecane (13.6 min), and *n*-tetradecane (15.2 min). Indeed, the cycloalkanes are just pointing out, as seen in-between the *n*-alkanes in the 83 window.

The extracted ion profiles presented in Figure 7.24, present a very clear pattern of aromatics that is matching the pattern shown in Figure 7.13.

The pattern shown by the sample was not generated solely from interfering products there are ILR contributing to this pattern. A very strong signal of that evidence is that the ratio of the methylnaphthalenes peaks at 14.0 and 14.3 min; 2-methylnaphthalene is greater than 1-methylnaphthalene, which is highly indicative of the presence of ILR.

The much greater aromatic content compared to the aliphatic content indicates that the pattern originates from weathered gasoline rather than a medium petroleum distillate.

Case 5

This sample is constituted of burned paper and tape that were removed from the room of origin of a house fire. Figure 7.25 shows the TIC of the extract. The chromatogram presents a familiar pattern, but there are a large number of extraneous peaks. Some of the greatest

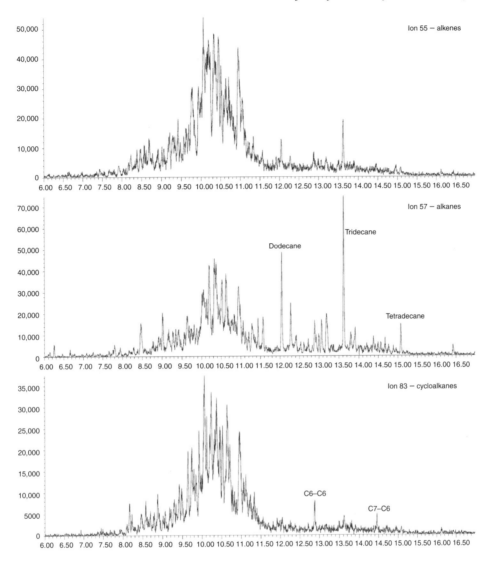

Figure 7.23 Extracted ion profiles of ions 55, 57, and 83 of the sample in Case 4

peaks are 1-butanol (3.4 min), propylene glycol (4.4 min), toluene (5.0 min), hexanal (5.3 min), furfural (5.8 min), ethylbenzene (6.2 min), *m*- and *p*-xylenes (6.3 min), *o*-xylene (6.7 min), 3-ethyltoluene (8.0 min), methylheptanol (8.3 min), 1,2,4-trimethylbenzene (8.6 min), cyclohexane (9.13 min), *d*-limonene (9.2 min), 3-furanmethanol (10.8 min), and two substituted propanoic acids at 14.7 and 15.0 min.

The presence of oxygenated compounds such as furfural, furanmethanol, butanol, hexanal, etc. is not surprising. Since the substrate is paper, it is expected to obtain these kinds of pyrolysis products. Also, there is tape present in the sample. The housing of the tape is probably made of polystyrene and the tape itself of polyester, as seen previously. This contributed

221

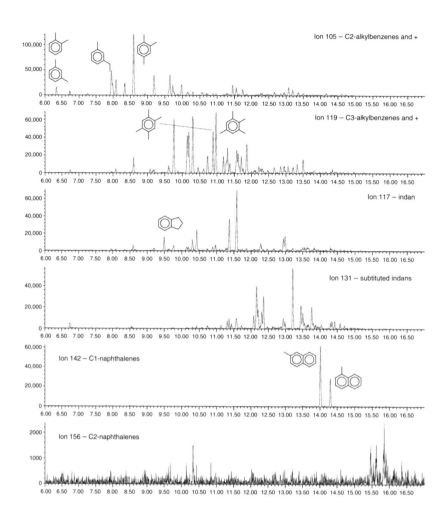

Figure 7.24 Extracted ion profiles of ions 105, 119, 117, 131, 142, and 156 of the sample in Case 4

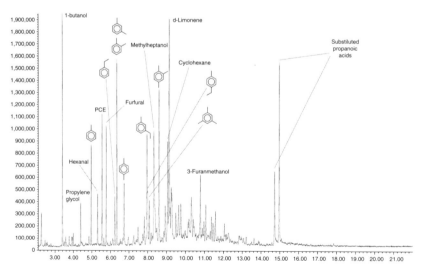

Figure 7.25 The TIC of the extract of a sample of burned paper and tape from a house fire (Case 5)

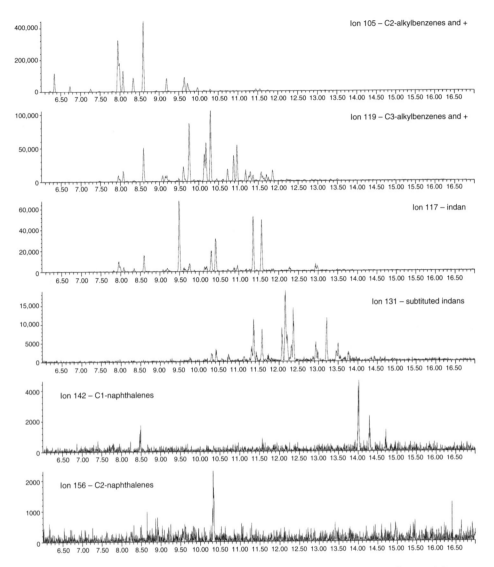

Figure 7.26 Extracted ion profiles for the ions 105, 119, 117, 131, 142, and 156 of the sample in Case 5

slightly to the pyrolysis products content, however styrene and α-methylstyrene are very small peaks.

By contrast the strong presence of certain aromatic compounds such as 1,2,4-trimethylbenzene, 1,3,5-trimethylbenzene and some other C2 and C3-alkylbenzenes, should catch the attention of the scientist and will require more investigation. Figure 7.26 shows the extracted ion profiles for the ions 105, 119, 117, 131, 142, and 156.

When compared to Figure 7.13, it is obvious that the pattern of all these ions is perfectly matching, with the exception of ion 156, in which the expected peaks are absent in the

223

sample. This is encountered with some gasolines. In this instance, the presence of weathered gasoline in the sample does not cause any doubt.

Conclusion

The complicated part of fire debris analysis is the interpretation of the chromatograms. This interpretation is difficult due to the facts that ignitable liquids encompass hundreds of different compounds, and that their patterns are modified by weathering and other degradation mechanisms occurring in the sample. Also, interfering products in the substrate that are analysed simultaneously with the compounds of interests can cause great complications.

These interfering products originate from three sources. First, there are the substrate background products (SBPs) that can be divided in natural products, manufacture induced products, and contamination (direct and indirect). Second, there are the pyrolysis products, produced by the burning polymers that comprise most common substrates. Finally, there are the combustion products, created by the oxidation of pyrolysis products and poor burning conditions.

Substrates submitted to fire debris analysts for ILR recovery and identification are complex in the sense that they are made of multiple different polymers, partially burned, and naturally contaminated by their everyday use.

When levels of ignitable liquids are high among a substrate, there is very little doubt about the identification of such ILR. However, when the levels get low enough to start being confused with the background, it becomes very difficult to render a conclusion.

By knowing the substances as well as the chemistry of the polymers that are used in everyday life and that are submitted to the laboratory as fire debris, it is possible to better understand the kind of products that are encountered in negative samples.

This chapter not only presented the basis of understanding such concepts, but it also provided a few examples of real samples typical of those encountered every day in the laboratory. Hopefully, the reader would have grasped the concepts explained and understood the examples. It would require several books such as this one to go over most of the substrates and to present every kind of possible situation. This is not the purpose of this chapter. However, it has presented enough of the basic concepts and ideas for the fire debris analyst to start studying 'blank' substrates with another view and to get more accustomed to these products.

Acknowledgements

The author would like to deeply thank Julia Dolan, Senior Forensic Chemist, Bureau of Alcohol, Tobacco, Firearms, and Explosives, Ammendale, Virginia, USA, for sharing her expertise and for her continuous and invaluable support, Reta Newman, Crime Laboratory Director, Pinellas County Forensic Laboratory, Largo, Florida, USA, for sharing her knowledge and for her devotion, as well as John Lentini, Fire Investigation Manager, Applied Technical Services, Marietta, Georgia, USA for his support and advice. The author also acknowledges that Figures 7.9 through 7.11 are results of research for a Master's thesis conducted at Florida International University, Miami, Florida.

References

1 D.M. Lucas (1960) The identification of petroleum products in forensic science by gas chromatography. *Journal of Forensic Sciences* 5(2): 236–247.

2 B.B. Coldwell (1957) The examination of exhibits in suspected arson cases. *Royal Canadian Mounted Police Quarterly* 23(2): 103–113.

3 D.L. Adams (1957) The extraction and identification of small amounts of accelerants from arson evidence. *The Journal of Criminal Law, Criminology and Police Science* 47(5): 593–596.

4 J.W. Brackett (1955) Separation of flammable material of petroleum origin from evidence submitted in cases involving fires and suspected arson. *The Journal of Criminal Law, Criminology and Police Science* 46(4): 554–561.

5 J.M. Macoun (1952) The detection and determination of small amounts of inflammable hydrocarbons in combustible materials. *The Analyst* 77: 381.

6 L.G. Farrell (1947) Reduced pressure distillation apparatus in police science technical notes and abstracts. *The Journal of Criminal Law and Criminology* 38(4): 438.

7 B.V. Ettling and M.F. Adams (1968) The study of accelerant residues in fire remains. *Journal of Forensic Sciences* 13(1): 76–89.

8 R.W. Clodfelter and E.E. Hueske (1976) A comparison of decomposition products from selected burned materials with common arson accelerants. *Journal of Forensic Sciences* 22(1): 116–118.

9 C.L. Thomas (1978) Arson debris control samples. *Fire and Arson Investigator* 28(3): 23–25.

10 M.R. Smith (1982) Arson analysis by mass chromatography. *Analytical Chemistry* 54(13): 1399A–1409A.

11 M.R. Smith (1983) Mass chromatographic analysis of arson accelerants. *Journal of Forensic Sciences* 28(2): 318–329.

12 J. Howard and A.B. McKague (1984) A fire investigation involving combustion of carpet material. *Journal of Forensic Sciences* 29(3): 919–922.

13 I.C. Stone and J.N. Lomonte (1984) False positive in analysis of fire debris. *The Fire and Arson Investigator* 34(3): 36–40.

14 J. Nowicki and C. Strock (1983) Comparison of fire debris analysis techniques. *Arson Analysis Newsletter* 7: 98–108.

15 J.D. DeHaan and K. Bonarius (1988) Pyrolysis products of structure fires. *Journal of the Forensic Science Society*, 28(5–6): 299–309.

16 W. Bertsch (1994) Volatiles from carpet: a source of frequent misinterpretation in arson analysis. *Journal of Chromatography A* 674: 329–333.

17 R.O. Keto (1995) GC/MS data interpretation for petroleum distillate identification in contaminated arson debris. *Journal of Forensic Sciences* 40(3): 412–423.

18 M.E. Kurz, S. Schultz, J. Griffith, K. Broadus, J. Sparks, G. Dabdoub, and J. Brock (1996) Effect of background interference on accelerant detection by canines. *Journal of Forensic Sciences* 41(5): 868–873.

19 D.J. Tranthim-Fryer and J.D. DeHaan (1997) Canine accelerant detectors and problems with carpet pyrolysis products. *Science and Justice* 37(1): 39–46.

20 J.J. Lentini (1998) Differentiation of asphalt and smoke condensates from liquid petroleum distillates using GC/MS. *Journal of Forensic Sciences* 43(1): 97–113.

21 J.J. Lentini, J.A. Dolan, and C. Cherry (2000) The petroleum-laced background. *Journal of Forensic Sciences* 45(5): 968–989.

22 K. Cavanagh, E. Du Pasquier, and C. Lennard (2000) Background interference from car carpets – the evidential value of petrol residues in cases of suspected vehicle arson. *Forensic Science International* 125: 22–36.

23 M. Fernandes, C. Lau, and W. Wong (2002) The effect of volatile residues in burnt household items on the detection of fire accelerants. *Science and Justice* 42(1): 7–15.

24 É. Stauffer (2001) Identification and characterization of interfering products in fire debris analysis, in International Forensic Research Institute, Department of Chemistry, Florida International University, Miami, FL.

25 W.M. Haessler (1974) *The Extinguishment of Fire*. National Fire Protection Association, Quincy, MA.

26 Massachusetts Chapter International Association of Arson Investigators (2000) *A Pocket Guide to Accelerant Evidence Collection*, 2nd edn, Brimfield, MA.

27 B. Chanson, E. Ertan, E. Du Pasquier, O. Delemont, and J.C. Martin (2000) Turpentine identification in fire debris analysis. in Second European Academy of Forensic Science Meeting, Cracow, Poland.

28 M. Higgins (1987) Turpentine, accelerant or natural ??? *The Fire and Arson Investigator*, 38(2): 10.

29 J.J. Lentini (2001) Persistence of floor coating solvents. *Journal of Forensic Sciences* 46(6): 1470–1473.

30 E. Locard (1920) L'enquête criminelle et les méthods scientifiques, Paris.

31 T. Lang and B.M. Dixon (2000) The possible contamination of fire scenes by the use of positive pressure ventilation fans. *Canadian Society of Forensic Sciences Journal* 33(2): 55–60.

32 M.P. Koussiafes (2002) Evaluation of fire scene contamination by using positive-pressure ventilation fans. *Forensic Science Communications* 4(4).

33 S.L. Madorsky (1964) Thermal degradation of organic polymers, in *Polymer reviews*, H.F. Mark and E.H. Immergut (ed.), vol. 7. John Wiley & Sons, New York.

34 T.P. Wampler (1995) Analytical pyrolysis: an overview, in *Applied Pyrolysis Handbook*, T.P. Wampler (ed.) Marcel Dekker, New York, pp. 1–29.

35 CDS Analytical, Inc, Degradation Mechanisms – Random Scission. 2000.

36 CDS Analytical, Inc, Degradation Mechanisms – Side Group Elimination. 2000.

37 CDS Analytical, Inc, Degradation Mechanisms – Depolymerization. 2000.

38 É. Stauffer (2003) Basic concept of pyrolysis for fire debris analysts. *Science and Justice*, 2003.

39 D. Drysdale (1985) *An Introduction to Fire Dynamics*, John Wiley and Sons, Chichester.

40 M.T. Pinorini (1992) La suie comme indicateur dans l'investigation des incendies, in Faculté de Droit, Institut de police scientifique et de criminologie. Université de Lausanne, Villard-Chamby.

41 M.T. Pinorini, G.J. Lennard, P. Margot, I. Dustin, and P. Furrer (1994) Soot as an indicator in fire investigations: physical and chemical analysis. *Journal of Forensic Sciences* 39(4): 933–973.

42 S.C. Moldoveanu (1998) *Analytical Pyrolysis of Natural Organic Polymers. Techniques and Instrumentation in Analytical Chemistry*, vol. 20. Elsevier, Amsterdam.

43 American Society for Testing and Materials (1997) ASTM E 1618–97 Standard guide for identification of ignitable liquid residues in extracts from fire debris samples by gas chromatography–mass spectrometry, in *Annual Book of ASTM Standards*, United States of America, pp. 654–659.

44 American Society for Testing and Materials (2002) ASTM E 1618-01 Standard test method for ignitable liquid residues in extracts from fire debris samples by gas chromatography–mass spectrometry, in *Annual Book of ASTM Standards, 2002*, United States of America.

45 S.R. Sandler *et al.* (1998) *Polymer Synthesis and Characterization*, Academic Press, San Diego.

46 University of Southern Mississippi (2002) The macrogalleria – a cyberwonderland of polymer fun. Department of Polymer Sciences.

47 American Society for Testing and Materials (2001) ASTM E 1412-00 Standard practice for separation of ignitable liquid residues from fire debris samples by passive headspace concentration with activated charcoal, in *Annual Book of ASTM Standards, 2001*, United States of America, pp. 431–433.

Index